THE DRAGON HUNTERS

ALSO BY FRANK GRAHAM, JR.

Illustrations by Robert Shetterly

T·T

A Truman Talley Book • E. P. DUTTON, INC.

THE
DRAGON
HUNTERS

Frank Graham, Jr.

Published in the United States by
Truman Talley Books • E. P. Dutton, Inc.,
2 Park Avenue, New York, N.Y. 10016

Library of Congress Cataloging in Publication Data

Graham, Frank, 1925–
The dragon hunters.
"A Truman Talley Book."
1. Insect control—Biological control—United States.
2. Insects, Injurious and beneficial—United States.
3. Agricultural pests—United States. 4. Pest control—
Biological control—United States. I. Title.
SB933.3.G75   1984   632′.96′0973   84-1586

ISBN: 0-525-24249-X

Published simultaneously in Canada by
Fitzhenry & Whiteside Limited, Toronto

Designed by Mark O'Connor

10 9 8 7 6 5 4 3 2 1

First Edition

TO THE MEMORY OF ROBERT VAN DEN BOSCH,
SCIENTIST AND CONSERVATIONIST

CONTENTS

CONTENTS

Contents

CONTENTS

THE DRAGON HUNTERS

INTRODUCTION

TWO DECADES HAVE PASSED SINCE RACHEL CARSON CON-
cluded in *Silent Spring* that "the chemical war is never
won, and all life is caught in its violent crossfire." Other
voices, equally urgent if less melodious, took up the cry at
the beginning of the 1970s and led us into the Age of Ecology, when pesticide dependency, as well as air and water
pollution, was expected to wither away like the state in a so-
cialist millennium. Environmentalists are as prone to stum-
ble over human nature as was Karl Marx.

The trouble was that the products Carson complained

of were new, but the practice was ingrained. Human beings, the only organisms with the wit to devise a religious and philosophical apologia for their appetites, discovered early on that their reach for dominion over earth and air was being challenged by the adaptable and durable—most ominously, the incredibly prolific—class of animals they called insects. The Hundred Years War pales into insignificance before the endless hostilities between the tiller/peasant/yeoman/ farmer/agronomist and the despised "bug." When push comes to shove in the trenches, the combatants turn to the most direct, if untidiest, means of slaughter.

From the earliest times, wherever humans sowed a field, we can be sure that other forms of life came to nibble at its edges. It might have been a deer or a rabbit (remember, Peter Rabbit was a pest in Mister McGregor's garden), but insects are the most ubiquitous of all non-microscopic animals and they must have soon earned the enmity of the men and women who tilled the land. Still, it is easy to overestimate the amount of damage insects have inflicted on agriculture. Farmers (like businessmen, fishermen, and football coaches) are inveterate grumblers, inclined to pessimism and an exaggeration of their current difficulties. Yesterday's peasant and today's agronomist have both invoked the specter of famine in the face of insect attack, but one of the world's most distinguished entomologists, Vincent G. Dethier, has put the threat in better perspective.

"Granted that there are new insects and that many native insects have transferred their attentions to cultivated plants, the evidence that insects compete *seriously* with us for *food* is unconvincing," Dethier wrote in his book, *Man's Plague?* "Weather, plant pathogens, and complex socioeconomic factors are the principal agents that threaten our food supply. The role of the insect is indeed becoming more dominant as agriculture evolves, but criteria other than famine must be employed to assess it. The black mount of the Third Horseman of the Apocalypse, Famine, rides in many guises, but he is not an arthropod."

Why, then, did the crisis in insecticide use arise in the first place? Despite warnings from scientists, environmentalists, and frightened neighbors about the dangers of chemical poisons, why has the farmer persisted in spreading them throughout the environment, with the advice and consent of industry and many government entomologists? In brief, why are these substances, so costly in both dollars and environmental health, routinely turned to in modern agriculture?

Domestic plants are generally coddled things, hybridized and civilized so that they are unable to survive outside the cultivated plot, prey to the diseases and insects that their wild relatives are not. Before agronomists developed disease- and weather-resistant varieties and new methods of cultivation and fertilization, both the individual plants and the harvest itself were puny by today's standards. But in the main, those plants survived insect attacks. Natural biological control has been going on since the beginning of time. Many potential pests were kept at reasonably low densities by their natural enemies—certain mammals and birds, as well as predatory and parasitic species among the insects themselves. Nature had learned to include within her protective apparatus even those plant varieties produced by upstart humans.

Colonists in the New World were sorely troubled by insects almost from the start. In many cases, however, the troubles were of their own importation, the colonists having brought pest insects with them from Europe in their seeds, straw, and other cargo. Free of their traditional natural enemies in their new home, these pests flourished. (A serious pest on the wheat crop in the new United States was called the Hessian fly because of the widespread though probably erroneous belief that it had been brought over in the straw bedding of the mercenaries who fought on the American side during the Revolution.) Yet agriculture prospered so extensively that, if England could be called a nation of shopkeepers, the United States could have been called, for a century or more, a nation of farmers.

In much of the rest of the world, agriculture was still being carried on by traditional methods. But nineteenth-century North America prefigured a new era. American agriculture became the wonder of the world for a host of reasons: vast expanses of cheap land, thousands of immigrants, eager for a new start, and by mid-century, a network of canals and railroads to bring farm products to the cities. Farm production soared as better equipment and new plant varieties became available.

But the seeds of trouble were there as well. Large farms and specialized equipment made it more profitable for the farmer to concentrate on a single crop. Into these monocultures rushed opportunistic insects, specialized to feed on a single crop, and thus the "superpests" were created. The farmer's plight was commented on shortly after the Civil War by a prominent entomologist, Benjamin Walsh, who had come to the United States after studying at Cambridge University.

"It is a remarkable fact that fully one-half of our worst insect foes are not native American citizens, but have been introduced here from Europe," Walsh wrote. "The plain common sense remedy for such a state of things is, by artificial means, to import the European parasites that in their own country prey upon the wheat midge, the Hessian fly, and other imported insects that afflict the North American farmer. Accident has furnished us with the trouble; science must furnish us with the remedy."

Here and there, farmers have always practiced biological control. Centuries ago, Chinese farmers put ants to work to control leaf-eating insects on their fruit trees, even building bridges of bamboo poles to help the ants cross from tree to tree. Perhaps the first case of "classical" biological control, which entomologists define as the importation of natural enemies to control a pest that is itself an alien, occurred in 1762. In that year, some farmers in India imported myna birds to feed on the red locusts that had become pests in their fields.

Biological control, in fact, may provide a simple key to even a bizarre dilemma. When Europeans settled Australia, which had few large native mammals, they imported cattle, horses, sheep, and other domestic livestock. As the settlers' herds increased, the animals littered the pastures with dung. On other continents, dung-eating beetles quickly clean up pastures, but in Australia, the dung lay where it fell because the native insects had evolved to deal only with the dry pellets of marsupials. The wet dung smothered the grasses and provided breeding places for pestiferous flies. Australian entomologists eventually solved their peculiar problem by bringing in beetles that feed on the dung of large herbivores from Africa. Even cow patties have their natural enemies.

But biological control, as we shall see, is a sophisticated science, requiring a great deal of skill, experience, and cooperation. A farmer, suddenly confronted with a serious pest, of whose natural history he may have no knowledge (and of whose name he may not even be certain), either endures the scourge with the legendary mute forbearance of his kind or strikes back as quickly and violently as possible. In the case of insects, the commonsense remedy seemed, to most farmers, to be a poison spread on or around crops to kill as many pests as possible.

One of the few practical poisons of the nineteenth cen-

Dung beetles gave Australia a unique form of bio control.

tury was arsenic. Two of the substances that came into favor as insecticides were known by the colorful cognomens Paris green and London purple, compounds originally marketed as dyes, but ingeniously turned against insects because they contained arsenic. Other popular insecticides of the time, delivered to the affected plants or fruit trees by rather primitive spraying contraptions, were kerosene emulsions and a mixture of sulfur, lime, and salt. Later, entomologists recommended calcium arsenate dust for some crops, especially cotton.

American entomology, when it came into its own, prized the dead insect, not simply as a specimen for study but also as a victim of a pest control program carried out for the benefit of the individual who was, in a very real sense, the entomologist's patron—the beleaguered farmer. In the United States, the entomologist rarely matched the image of his European counterpart—the gentleman collector netting butterflies in a Alpine meadow or the scholarly J. Henri Fabre endlessly observing *Sphex* wasps on a parched plot in Provence. American entomologists came to view the killing of bugs as their highest mission.

"These people *loathe* insects," an entomologist said some years ago, referring to many of his colleagues. "Their life is a crusade against them!"

By the beginning of the twentieth century, much American farming, based on extensive monocultures and the consequent huge capital investments in land and machinery, had become big business. Often burdened with pressing debts because of this investment, the farmer could not tolerate crop failures, or what he considered to be unacceptable damage by insects. Farmer organizations had considerable influence in Washington, so that eventually the United States Department of Agriculture (USDA) and its counterpart in the individual states became largely promotional and protective arms of the agriculture industry. They offered what was almost the only secure employment for the country's early entomologists.

"Establishment of the land grant universities, state agricultural experiment stations, and extension services created additional professional opportunities in each state of the union," wrote John H. Perkins, who traced the development of economic entomology in the United States in his book *Insects, Experts, and the Insecticide Crisis.* "These positions, combined with those of the federal government, became the major market place in which entomologists could sell their knowledge to farmers. Entomology passed from an amateur's hobby to full-time professional work. By the beginning of the twentieth century, it was clearly possible for a young man (occasionally a young woman) to contemplate making a living by studying insects and their control. Development of a large insecticide industry later served as a supplementary market for entomological expertise, but these developments were confined largely to the years after 1945."

The year 1945 can be seen as the great continental divide of insecticide use. Entomologists, through the universities or the government agencies for which they worked, were almost entirely beholden to the agricultural establishment. Some of them, it is true, tried to follow Benjamin Walsh's dictum that "science must furnish the remedy" to insect pest problems, and they advised farmers on a variety of cultural methods, such as crop rotation, early harvests, and stubble removal to alleviate the threat of insect damage. But insecticides produced effects that could be observed (with satisfaction) by the farmer. Poisons allowed the farmer to watch stricken bugs spinning out their lives in the dust. What the total effect on the environment was, or how much the farmer actually saved in dollars and cents by the use of poisons, were questions that did not always bear looking into. Entomologists learned that the surest way to satisfy the farmer was to provide him with the sight of thousands of dead insects. In the years leading up to World War II, entomologists and the growing American chemical industry added to the farmers' chemical arsenal such chemicals as ethylene oxide and paradichlorobenzene. Also falling into place was the apparatus for delivering those chemicals to the field, including batteries of salesmen, more effective spraying equipment, and crop-dusting aircraft. Chemicals overshadowed all other methods for dealing with pest insects.

"The failure of other methods to meet public demands for ways to stop insects without long, expensive research, changes in farming practices, or long-term planning paved the way for chemicals," writes Thomas R. Dunlap in *DDT: Scientists, Citizens, and Public Policy.* "The triumph of chemical insecticides was due not just to the visible results they gave, but to their acceptance by a public and a farming community that valued, above all else, convenience, simplicity, and immediate applicability. That economic entomologists recognized that need and were prepared to meet it can be seen not only in the complaints that the entomologist was 'losing sight of the insect,' but in positive exhortations to use insecticides. . . . Profit was the most tangible and probably the strongest motive. Alone among insect remedies, chemicals lent themselves to quick, large-scale commercial exploitation, and by 1910 there were so many manufacturers, and there was so much fraud, that Congress passed the Insecticide Act of 1910, which established standards and a regulatory apparatus. As early as 1912 an entomologist in California complained of the 'tremendous influence the manufacturers and dealers of insecticides are exerting. Their salesmen see more farmers than the county agents can, they give the last advice before the farmer applies the product, and they can counteract our recommendation.' By the 1920's [USDA's Bureau of Entomology] was working closely with commercial interests in the field and the partnership continued to grow."

Out of the technology of World War II came a flood of new chemicals that promised to enhance human life all over the globe. Among the most glamorous was the insecticide DDT (dichlorodiphenyltrichloroethane) and such chemical relatives as aldrin and dieldrin. A pest control specialist's delight, DDT was extremely toxic to insects, remaining effective for long periods of time in the open environment and yet seeming to pose no threat, at recommended levels of application, to humans and other vertebrate animals. There were some doubting Thomases from the beginning because no one had really tested DDT's safety in the field, but the view of economic entomologists and industry officials prevailed.

"Is this not an auspicious time," asked an influential entomologist in 1946, "for entomologists to launch determined campaigns for the complete extermination of some of the pests which have plagued man through the ages?"

There no longer seemed any limits to man's complete domination of nature, at least down on the farm. But signs of trouble were apparent as early as the 1950s. The precipitous decline in number of great birds of prey caused a few researchers to take a closer look at DDT and the pathways through which its residues moved up through food chains to strike at the reproductive systems of such top carnivores as the bald eagle, the osprey, and the peregrine falcon. Pest insects acquired resistance to the new chemicals and evaded control measures. Other insects, which had never before been considered troublesome to humans, suddenly increased in number to become pests, as the predators and parasites that had once controlled them proved especially vulnerable to the new insecticides. Serious questions arose about the effects of DDT on human health. Euphoria turned to disillusion.

One of the scientists who grew alarmed at these developments was Rachel Carson, a marine biologist and the author of several beautifully written books about the sea. She spent four years looking into the subject, not an easy job by any means because industry and the USDA presented a monolithic front against anyone who challenged a pest control strategy based on the profitable chemicals. Wildlife officials, having observed the disasters triggered by pesticide misuse, were eager to help, as were some physicians, farmers, and environmentalists. Even within agricultural agencies, there were "moles" who uncovered the kind of information Carson needed for her book. One of them was Reece I. Sailer, later Graduate Research Professor at the University of Florida, but for many years a specialist in biological control for the USDA.

"There were two years in which I accomplished more for biological control than during the rest of my career," Sailer told a meeting of environmentalists in Washington, D.C., in 1976. "That was while I was engaged in an almost clandestine activity, assisting Rachel Carson in obtaining bibliographical materials for *Silent Spring.* I did not let even my colleagues in the old entomology division know too much of what I was doing. They might wonder who came into Sailer's office, but they did not know her, so there was no problem."

Carson's *Silent Spring* was published amid a great uproar in 1962. The counterattack from those who supported the unrestricted use of chemical insecticides proved to be more savage than even the apprehensive author had anticipated. The chemical industry and the nutrition foundations it supported distributed "fact kits," disputing Carson's revelations on the environmental problems chemical insecticides had already caused and the threats they posed for the future. Their attitude seemed to be that, although Carson had called for a ban only on DDT and the other extremely persistent insecticides that contaminated food chains, her indictment was comprehensive enough to cause questions to be asked about all insecticides. Typical of the level of attacks mounted against *Silent Spring* was that by the director of the New Jersey Department of Agriculture, who wrote: "In any large scale pest control program we are immediately confronted with the objection of a vociferous, misinformed group of nature-balancing, organic-gardening, bird-loving, unreasonable citizenry that has not been convinced of the important place of agricultural chemicals in our economy."

Silent Spring survived the attacks to come down in history as one of the century's landmark books. When President Kennedy appointed a commission to look into pesticide use in the United States, its findings substantiated most of Rachel Carson's claims. (Carson died in 1964.) As evidence against its environmental effects piled up, DDT came under increasing attack. The issue was finally joined during a hearing in Wisconsin in 1968–69 before the Department of Natural Resources to consider a citizens' petition to ban DDT. The scientific testimony assembled there by the Environmental Defense Fund destroyed the credibility of DDT's defenders and led to a nationwide ban on that chemical, for most purposes, by the Environmental Protection Agency in 1972. Aldrin, dieldrin, and most of the other related persistent chemicals came under a similar ban in the next few years. Consequently, levels of DDT in human beings, as well as in many birds and mammals, have now declined, and species such as the bald eagle and the osprey have shown an encouraging, if somewhat slow, recovery in number.

Although DDT has been almost completely phased out in the United States, Canada, and in most other developed nations, many other dangerous pesticides are still produced in large quantities for a variety of uses, and in increasing amounts. "The current situation is a national scandal," *The Amicus Journal* (the publication of the Natural Resources Defense Council) reported in its Summer, 1983 issue. "In the past thirty years, the use of pesticides has increased nearly ten-fold (from 200 million pounds in 1950 to more than 1.7 billion pounds in 1980); and the overwhelming irony is that American farmers lose twice as many crops to pests as before the big push to chemicals. The reason: the bugs simply become immune."

Yet juggling statistics when talking about pesticides is a tricky, and often misleading, game. Records are kept indifferently in many parts of the country, and comparing usage in dollars or pounds from one decade to another can be like comparing apples and oranges; inflation, or the turn to other, more sophisticated chemicals, makes gibberish of the numbers. It is true that there has been an enormous increase in the quantities of pesticides used in American agriculture during the last three decades, but because the figures are not broken down by type of pesticide, they are difficult to interpret. In fact, the bulk of the increase in pesticide use can be attributed to agriculture's growing dependence on herbicides to control weeds. The increase in insecticides, used to kill insects, has not risen at a comparable rate, if at all, in recent years.

Government agencies often have only a hazy idea of where the chemicals bought by growers are applied, and in what amounts. In Texas, where the figures are generally dependable, an observer who confined his reading only to cotton statistics might conclude that insecticides are on the way out. There, as we shall see, an imaginative integrated pest management program cut the use of insecticides on cotton fields from over 20 million pounds in 1964 to about 2 million pounds in 1984. But that decline is in part offset by the sorghum crop, which accounted for 200,000 pounds of insecticides in 1964 and then surged to more than 4 million pounds by the late 1970s.

Differences in pounds applied may also be deceptive. Some of the insecticides phased out as environmentally harmful because of their persistence and magnification through animal food chains were not acutely toxic, and thus growers applied them (often in the form of heavy dusts or powders) to their fields in staggering amounts. They deliver the newer, "hotter" (or extremely toxic) chemicals in fine sprays. Poundage is down, but toxicity may very well be up.

Inevitably, the environmental costs are severe. Scientists at Cornell University have placed the number of nonfatal human cases of pesticide poisoning at more than 100,000 a year, whereas a government survey of hospital records in the United States disclosed an annual average of 65 pesticide-related deaths. Despite conflicting opinions, no one has yet determined the long-term effects of these chemicals on humans.

Total wildlife losses, too, are anyone's guess, although there has been a concerted effort at least to gather statistics on the destruction of the economically indispensable honeybees, which pollinate more than 1 billion dollars worth of agricultural crops each year. A survey of California's one-half million honeybee colonies in the 1970s demonstrated a 13 percent average annual loss to chemicals, with consequent untold decreases in yields to fruit and vegetable growers, as well as a direct cost to the beekeepers themselves.

There is no question that there are still legitimate needs for these chemicals on crops that come under ruinous insect attack. It would prove a financial hardship to certain growers to withdraw insecticides all at once, and in other places, these chemicals play a valuable role in integrated pest management programs. But as Paul R. Ehrlich, Professor of Biological Sciences at Stanford University, has written, insecticides are like heroin in that "they promise paradise and deliver addiction."

Yet much use of chemical insecticides has nothing whatever to do with famine or even undue financial hardship. As we shall see, a large proportion of the insecticides sold in the United States is for use on cotton; it is difficult to justify this use to kill insects that may take a part of the cotton crop when the federal government pays growers millions of dollars each year to keep cotton fields out of production. Another substantial proportion is applied not to save crops from destruction, but merely for cosmetic purposes; much chemical spraying is directed at insects that cause minor discoloration or scuffing of the skins of certain fruits and vegetables. For instance, chemicals that are possible carcinogens are applied to oranges to protect them from an innocuous insect that roughens parts of their skins. If Rachel Carson were alive today, she would not have much reason to change her view that we are still mired in "the Stone Age of Science." Her plea was that we reduce our dependence on those substances that have brought us no closer to solving our pest insect problems, and turn instead to biological control. Although impressive advances have been made in this field in recent decades, its acceptance in many parts of the country continues to be painfully slow. A decade ago, Reece Sailer calculated that the chemical industry spent at least 110 million dollars a year on the research and development of pesticides, whereas since 1888 (!), the total spent by the federal government and all the states for the introduction of natural enemies amounted to about 20 million dollars. The gap has only widened in the meantime.

The following chapters demonstrate that biological control is not an ivory tower exercise, but rather a viable alternative for men and women in a business in which a few extra dollars in operating costs may tip the balance from success to failure. A survey by entomologists at the University of California, Riverside, showed that, easily, the highest cultural operating cost for citrus growers in parts of Ventura County in 1980 was for pest control (exceeding that for irrigation, fertilization, weed control, pruning, frost protection, and other day-to-day needs). That cost amounted to 375 dollars an acre. By 1982, after the introduction of various natural enemies of the insects and a cutback in spraying, the pest control cost had dropped to 160 dollars an acre.

Paul DeBach, of the University of California, Riverside, has pointed out that the introduction of a natural enemy may cost anywhere from less than 100 dollars to, occasionally, tens of thousands of dollars, paltry in any case when compared with the investment made in chemicals. DeBach concluded that the use of chemicals saves growers five dollars for every one spent; the savings from reduced insect damage and the costs of chemical treatment saved growers 30 dollars for each dollar spent on the importation and colonization of natural enemies. Biological control has no reason to exist unless it enables the grower to make a profit.

But, some skeptics have asked, might we not be letting ourselves in for a nasty surprise with this method of pest control, as we have with chemical control? What's to prevent an insect, imported from abroad to kill pests, from eventually turning into a pest itself? The answer (to be discussed in detail in the following chapters) is that these natural enemies are highly adapted to the insects they prey upon. They have evolved with them through the ages, adapting to overcome the defenses of their particular victim. No insect is released in biological control until specialists have determined its binding ties to the pest it was imported to fight. In a sense, these natural enemies are the sum of the adaptations they have made to deal with their victim—the length and strength of the ovipositor needed to implant an egg in its body, the ability to detect the prey's distinctive chemical cues, the inherited characteristics that time their reproductive cycles to the developing stages of the pest.

A good case can be made for the claim that applied biological control, which is the science of fighting pest insects and plants with other insects and sometimes with microorganisms, is today's most exciting and satisfying venture into natural history. Its practitioners ransack the far places of the earth to find the "wild animals"—the predators and parasites—that are the tools of their profession. They must bring them back alive; determine whether or not it is safe to release them into a new environment; calculate their chances of success against what is often an explosive and mobile population of insects that are considered harmful to mankind's best interests; rear them in numbers sufficient to secure their success; and see that they are able to establish themselves in a land with which they are completely unfamiliar. Someone has aptly called this science "the thinking person's pest control."

1

FIRST VICTORY

The Cottony-Cushion Scale
and the Ladybug

THE COTTONY-CUSHION SCALE (*ICERYA PURCHASI*), LIKE SOME unsavory but ill-starred felon, enjoyed a brief notoriety in the American press before dropping back into relative obscurity. Its fling was virulent, its comeuppance astonishingly abrupt. Old biological control men speak of it almost fondly, in the manner of retired detectives recalling how a young punk's criminal ambitions were nipped in the bud by imaginative police work.

Scale insects are close relatives of the mealybugs and rather more distant ones of the aphids. Human aversion to them stems, in part, from their general untidiness; many species live together in colonies on the tender leaves they suck dry, and spatter the surface with their exudates of wax or honeydew. Most of them, such as the cottony-cushion scale, injure plants wherever they are not controlled. Other species have been of great value to mankind. One yields cochineal, the scarlet substance once widely used in dyes, cosmetics, and medicines. (The words for two of our commonest bright reds are associated with these insects—vermilion, through the Latin *vermiculus*, the "little worm" or scale from which a red dyestuff was extracted; and crimson, through *Kermes*, a genus of scale that infested the kermes oak and gave a dye that the Spanish called *cremesin*.) A species found in India, the lac insect (*Laccifer lacca*), used to be harvested in astronomical numbers to produce shellacs, varnishes, and linoleum. Scale insects secrete wax, from which the Chinese made candles, and honeydew, which is said to have been the manna of the Israelites. But synthetic substances have, in most cases, replaced the insect by-products, and the family Coccidae languishes in ill repute.

The armored scale insects are very small, flattened organisms that, as adults, live under their waxy cover. In some species, this cover takes the form of a shell-like scale that gives a colony on a tree the look of tiny limpets on a rocky shore. The cottony-cushion scale, which is about three-tenths of an inch long, forms a cottony, waxy coating marked by longitudinal ridges that provide the species with an alternate name, the fluted scale.

Although there is a wide variety of forms in the family, entomologists speak of the Coccidae as "degenerate." Adult females most closely fit the characterization, for they are wingless and often lack both legs and eyes. They are simply sap-sucking, egg-laying machines, and largely immobile.

They lay their eggs under their cover; some species promptly die and leave the eggs to shelter from the winter there.

Males, although they may grow wings, at times seem almost superfluous. They have no functional mouthparts for feeding and may take only occasional roles, or even no role at all, in the reproductive process, since the females often reproduce by parthenogenesis. The young, when they hatch, possess legs and move about before settling down to the largely sedentary life of an adult female or the brief sexual fling of the winged male.

Promptly at nine-thirty on Tuesday morning, April 12, 1887, Charles Valentine Riley rose to address the semiannual meeting of the California State Board of Horticulture at Riverside. The president of the board, Elwood Cooper, had introduced Riley as the chief of the U.S. Department of Agriculture's Division of Entomology and one of the country's foremost experts in devising ways and means to combat destructive insects. Fruit growers from all over southern California had assembled at the Riverside Pavilion to hear Riley's address, which the *Pacific Rural Press* was to describe some days later as "full of valuable information, and which, though quite lengthy, was listened to with the most marked attention throughout."

The fruit growers' concentration on Riley's address was not surprising. His subject was the cottony-cushion scale, alias the fluted scale, alias the white scale, a voracious pest that had descended on California as if out of nowhere and devastated many of the state's citrus groves. No serious remedy for it seemed to exist. It had already driven a number of growers out of business, while others predicted the imminent doom of an industry that was still comparatively young.

"Historical evidence all points to Australia as the original home of this insect, and its introduction from Australia to New Zealand, Cape Town, South Africa, and California," Riley told his audience. "Nothing was known or published upon the species prior to the seventh decade of this century, and it seems to have first attracted attention almost simultaneously in Australasia, Africa, and America. The evidence as to whether it is indigenous to Australia or New Zealand, or to both, is not yet satisfactory. The first personal knowledge which I had of it was from specimens sent to me in 1872 by Mr. R. H. Stretch, then living in San Francisco, and all the evidence points to its introduction into California by the late George Gordon, of Menlo Park, about the year 1868, and probably from Australia, on *Acacia latifolia.*"

Riley touched briefly on the natural history of scale insects, emphasizing the mobility of the cottony-cushion scale, compared to most of its relatives. The immature stages, he said, are especially active soon after hatching and because of their extreme lightness often travel considerable distances on the wind. Turning to remedies, he confirmed the general suspicion that there were no natural enemies of the pest, to speak of, in California and said that the most promising strategy yet tried was the application of a spray, a kerosene emulsion diluted with eight to ten parts of water; although Riley's agents had used the mixture on orange trees with promising results, commerical growers had found it less successful.

And then Riley delivered what was to be the most important part of his message. He recommended that the state, or even Los Angeles County, appropriate "a couple of thousand dollars" to send an entomologist to Australia to look for parasites of the cottony-cushion scale. As he pointed out, the scale seemed to have originated in that part of the world, but was seldom spoken of in Australia as a serious pest. Obviously, some natural force kept the scale population in check there.

"I would not hesitate, as United States Entomologist, to send someone there with the consent of the Commissioner of Agriculture, were the means for the purpose at my command," Riley said. "But unfortunately, the mere suggestion that I wanted $1,500 or $2,000 for such a purpose would be more apt to cause laughter and ridicule on the part of the average committee in Congress than serious and earnest consideration, and the action of the last Congress has rendered any such work impossible by limiting investigation to the United States."

The last Congress, indeed, had even limited Professor Riley, who had been in the habit of visiting Europe in search of natural enemies and charging his trips to the federal government. Many legislators looked on such ventures as the ultimate wild goose chase. Members of the California legislature apparently shared the skepticism of Congress in the matter of fighting insect pests with other insects, for upon his return to Washington, Riley learned that the fruit growers' plea for state funds had been rejected. But Riley was an aggressive and resourceful man. When he heard that the State Department was sending a delegation to an international exposition in Melbourne in 1888, he convinced the Secretary of State to set aside 2,000 dollars for an entomologist to travel as a member of the delegation.

Riley had no difficulty in choosing a "bug hunter" for the assignment. Albert Koebele, born in Germany, was a naturalized citizen of the United States and an experienced entomologist. He had come to Riley's attention some years earlier at a meeting of the Brooklyn Entomological Society.

"Riley had been impressed by the beautiful mounting

of specimens that Koebele exhibited in a collection and invited Koebele to come to Washington D.C., and work for Riley at the Division of Entomology," Richard L. Doutt wrote in *Biological Control of Insect Pests and Weeds* (edited by Paul DeBach). "Koebele seems to have been so overwhelmed by this invitation that he immediately gave up his job and got ready to go to Washington. There appears to have been several agonizing months of waiting, and finally in October 1881 he received a letter from Riley saying: 'I have felt aggrieved at having been the unintentional instrument in causing you to give up your former position before I had secured one for you in Washington.' Riley then offered him a position as a temporary agent at $65 per month, and added: 'I hope you will bring along all the specimens possible.' "

The following year Koebele entrenched himself in the Division of Entomology by taking part in an expedition to Brazil to investigate insects that were pests on cotton, sugar cane, and orange trees. Back in Washington, he became enmeshed in an unhappy affair of the heart and asked to be dispatched to some distant place. Riley, having been petitioned for help by California's orange growers brought to their wits' end by the depredations of the cottony-cushion scale, assigned Koebele to that and other insect control projects. Koebele worked closely in California with another of Riley's field agents, a former schoolteacher, D. W. Coquillett, who had contracted tuberculosis in Illinois and moved west to regain his health. The two men did not always agree in their methods, and after Koebele complained to Riley, Coquillett was even laid off for a while. Koebele prospered, however, having married and had his salary raised to $1,500 a year, and he was conveniently situated to leave for Australia with the State Department delegation.

What was to become the most celebrated of all foreign

explorations for beneficial insects began on August 25, 1888, when Koebele left San Francisco by steamship for Australia. After a voyage of three weeks he landed at Auckland, New Zealand, for a day's layover, which he spent diligently hunting for the cottony-cushion scale; he found only four full-grown females. Five days later he was in Sydney.

The trip went well from the start. Other collectors during the early days of foreign exploration were to encounter all sorts of frustrations and inconveniences in remote places, but Koebele's trip was eased by his status as a member of a prestigious diplomatic delegation. Australian entomologists and administrators extended him every courtesy. He traveled throughout the country on a free railway pass obtained for him in Sydney by the United States consul. Going on to Melbourne, he was welcomed by Baron Ferdinand von Mueller, the former director of the Botanic Gardens, who assured him that the scale "never became extensively injurious in Australia."

Riley's instructions to Koebele had emphasized the investigation of parasites. Just before Koebele had left San Francisco, a shipment of live parasitic flies, *Cryptochaetum iceryae,* had been received in the United States, sent by an Australian entomologist who had discovered them attacking the scale near Adelaide. Riley tended to discount this insect as a serious natural enemy because no dipterous fly had ever been recorded as parasitizing scale insects. (Riley was wrong, for *Cryptochaetum* first established itself in San Mateo County and remains a valuable natural enemy on the scale in southern California.) In any case, Koebele began to concentrate on parasites.

Adelaide, the site of *Cryptochaetum's* discovery, seemed to him the most likely place to begin his serious collecting. (He *was* a serious collector, later explaining away his failure to study the life histories of natural enemies in the field by citing his motto, "Get as many as possible.") He may have been disappointed in the beginning because hardly any of the gardeners he met at Adelaide had ever heard of the cottony-cushion scale, and an experienced museum entomologist there reported he had seen that insect for the first time only two years earlier. But on October 15, two entomologists escorted him to North Adelaide, where the scale was said to exist.

"We found there in one garden a few orange and lemon trees with the scales, which were subsequently collected for shipment," Koebele wrote in his official report. "In another garden, and also on orange, an occasional specimen was found. I discovered there, for the first time, feeding upon a large female Icerya, the Lady-bird which will become famed in the United States."

Koebele, of course, wrote that sentence from hindsight, many months later in California, very much aware by then of his momentous discovery. But at the time, when he called the attention of the two Australian entomologists to the small spotted beetle, both told him that they had never seen it before. The ladybird was to become known as the vedalia beetle (*Rodolia cardinalis*). Koebele collected the ladybird and went on to Mannum on the Murray River. There in an orange grove, he found larvae of the vedalia beetle feeding on an infestation of the cottony-cushion scale. He also found the parasitic fly *Cryptochaetum*, busily devouring the scale. He collected all the scale insects he could find, along with their parasites and predators, and packed them in boxes that he kept in a cool cellar in Adelaide until he could arrange to ship them to California.

"I finished collecting for my first shipment on the 25th and estimated that I had about 6,000 Icerya [cottony-cushion scales] which in turn would produce an aver-

Vedalia beetles prey on the cottony-cushion scale.

age of about four parasites each. They were packed partly in wooden and partly in tin boxes. Small branches generally full of scales were cut so as to fit exactly lengthwise into the box. With these the boxes were filled and all loose scales placed in between, plenty of space remaining for any of the insects within to move about freely without danger of being crushed by loose sticks. Salicylic acid was used in small quantities in the tin boxes to prevent mold, yet these, as I have been informed by Mr. Coquillett, arrived in a more or less moldy condition, while those in the wooden boxes always arrived safe."

On the two-day journey by train from Adelaide to Sydney, Koebele kept his shipment in an icebox aboard the sleeping car. At Sydney, he arranged with steamship officials to have the boxes kept in the ship's icehouse on the voyage to San Francisco, during which the temperatures would range from 38° F on leaving Sydney to about 46° F on arrival. Koebele experienced a moment of horror upon seeing the rough treatment accorded his precious boxes by deckhands in Sydney, but all the insects proved hardy. They carried on their life cycles and Coquillett told him later that some parasites were just emerging from their hosts upon arrival in California. Koebele remained behind, continuing to collect the scale with its parasites and predators through the antipodal summer. Only one of his shipments came to grief, when the boxes fell off the icehouse shelves during a gale and were crushed by falling cakes of ice.

The trip continued to go smoothly for Koebele, aside from the almost unbearable heat in southern Australia, where the temperature reached 108° F, and an incident in Melbourne, when, tempted to enter a churchyard to collect the scale, he was warned by a policeman "not to jump from the fence as they would surely have me arrested." Having concentrated on the parasite *Cryptochaetum* for the most part in Australia, he left there at the end of January 1889, and on Riley's orders, visited New Zealand again for a more thorough investigation of the scale there. He observed an outbreak of scale in one region, finding that the vedalia beetle had somehow arrived in its wake from Australia and was feeding on the pest.

"They were in such numbers that I found it not very difficult to collect here about six thousand specimens during the three days," Koebele wrote of the vedalia beetles. "As many as eight eggs of the Lady-bird were observed on the upper side of the female *Icerya* just beginning to exude cottony matter. Opposite to this on the small branch of Acacia, five young larvae of the Lady-bird were feeding on the under side of a half-grown scale; in one instance even nine larvae were found attached to a small *Icerya*. The mature beetles were not numerous, but every branch full of scales had a greater or less number of eggs and larvae. The eggs are chiefly deposited among the vigorous half-grown scales. Here the largest number of the eggs and young larvae were found. They are generally single, thrust in between the scales and fastened onto the branch, on the scale itself, and often on the underside of the scale, as the mother Lady-bird will sometimes raise the *Icerya* with her hind legs and thrust the egg under it."

Koebele packed his collection of vedalia beetles and, on February 25, left New Zealand for the voyage home.

"Having made arrangements with the butcher on board the previous day as to the most convenient time of receiving my insects into the ice house, they were transferred from the freezing-house on board the steamer which did not take more than ten minutes, and the insects were not disturbed in their dormant stage during the time," Koebele wrote. "Every day on the voyage I received the answer from the butcher, to my inquiries about the parcel, 'Your bugs are all right.' On March 10, after leaving Honolulu, one of the boxes with the Lady-bird larvae was examined and found in excellent condition; no dead larvae could be found among them, and this was twenty-four days after the first were collected. On Saturday evening, March 16, we arrived at San Francisco, too late to have the insects forwarded, and I could not send them off before Monday evening, March 18. They were probably received and opened by Mr. Coquillett two days later. This would make thirty-four days that they were enclosed, and yet they arrived in excellent condition, better than any previously received. Having been on ice for twenty-nine days, no doubt many of the eggs arrived here before hatching, and the larvae under such conditions would make little progress in their growth."

Besides his shipment of beneficial insects, Koebele returned with a great deal of information on the cottony-cushion scale. His talks with entomologists at various points on his trip had convinced him that the scale was indigenous to Australia and had spread from there to New Zealand, South Africa, and the United States. And his talks with pursers of the various steamships on which he sailed had convinced him of the ease with which the scale could inadvertently be spread from one continent to another. On almost every voyage, the ships carried many plants in their cargo, among which the scale could pass unobserved. For instance, during one season in the early 1870s, the markets in San Francisco were supplied entirely with Australian oranges.

Meanwhile, Koebele had at last found time to begin to know the vedalia beetle itself. A member of the family

Coccinellidae, it is one of the few kinds of insects that human beings have taken to their hearts. This affection goes back as far as the Middle Ages, when hard-pressed humans were likely to consider any beneficent aspect of the generally forbidding natural world to be heaven-sent. Rural folk noticed how efficiently these brightly colored, dark-spotted beetles attacked the hordes of aphids and other pest insects in their fields. To these people, they were "Our Ladie's beetles," tangible reflections of the blessings that stemmed from the Virgin Mary. The various vernacular names by which the insect is known—ladybird beetle, ladybug—refer to this. In fact, ladybirds seem to have few enemies anywhere. Bright coloring among insects is frequently an indication of inedibility, a warning to other predators that the creature in question is either poisonous or foul-tasting. Ladybird beetles, so it is said, have a strong, unpleasant taste and an equally unpleasant odor. The human impulse to squash every creature that crawls has been tempered in this case by the old rhyme, "Ladybird, ladybird fly away home, your house is on fire and your children all gone." The rhyme may refer to the practice of burning hop vines after the harvest in an attempt to kill the pest insects that wintered in them; the ladybird's larvae, or "children," were likely to be consumed in the conflagration. (*The Oxford Dictionary of Nursery Rhymes* notes that "the rhyme is undoubtedly a relic of something once possessed of an awful significance.") Although ladybirds sustained some losses there, the rhyme must have stayed many a child's hand that might otherwise have snuffed out the life of the insect.

Like any other family, ladybirds harbor a few undesirables. Several species, including the Mexican bean beetle, are plant-eaters and can become major pests. But most species of ladybirds feed on other insects. The vedalia beetle stands high among this latter group, its efficiency enhanced by both its voraciousness and fecundity.

"The period from the laying of the eggs until the adults again appear occupies less than thirty days for the vedalia," wrote C. V. Riley. "At this rate of increase calculating that three hundred eggs are laid by each female, and that one-half of these produce females, it will readily be seen that in six months the offspring of a single female beetle may under favorable circumstances amount to twenty-two trillion."

Coquillett began to receive shipments of vedalia beetles from Koebele in the late fall of 1888. He arranged with a property-owner in Los Angeles to release the beetles in a scale-infested orange tree over which a net had been erected. By spring, the ladybirds had multiplied prodigiously. Coquillett opened one side of the tent, permitting some of the hungry insects to escape to other trees in the vicinity, for they had obliterated the scale on the original tree. He also distributed the ladybirds to more than 200 growers in the region. The ladybirds prospered far beyond what anyone had imagined earlier. They went about their deadly work single-mindedly, moving from orchard to orchard and leaving behind the dried husks of scale insects still clinging to the trees by their beaks. In July 1889, J. R. Dobbins, a citrus grower of San Gabriel, testified to the phenomenon.

"The vedalia has multiplied in numbers and spread so rapidly that every one of my 3,200 orchard trees is literally swarming with them," he wrote. "All of my ornamental trees, shrubs, and vines which were infested with white scale, are practically cleansed by this wonderful parasite. About one month since I made a public statement that my orchard would be free from *Icerya* by November 1, but the work has gone on with such amazing speed and thoroughness that I am today confident that the pest will have been exterminated from my trees by the middle of August. People are coming here daily, and by placing infested branches upon the ground beneath the trees for two hours, can secure colonies of thousands of vedalia which are there in countless numbers seeking food. Over 50,000 have been taken away to other orchards during the past week, and there are millions still remaining, and I have distributed a total of 63,-000 since June 1. I have a list of 130 names of persons who have taken colonies, and as they have been placed in orchards extending from South Pasadena to Azusa, over a belt of country ten miles long and six or seven in width, I feel positive from my own experience, that the entire valley will be practically free from *Icerya* before the advent of the New Year.''

Other growers echoed Dobbins' wonderment, for they had been pulled back from the brink of bankruptcy. People flocked to local horticultural agencies to collect vedalia in pillboxes, spool-cotton boxes, and almost any small container they could find. Within a year or two, the cottony-cushion scale was only an uncomfortable memory. The infant orange-growing industry had been saved in California, and although there have been other scares from pest insects in subsequent years, a more mature industry was in a stronger position to deal with them. The cottony-cushion scale remains in California at an extremely low density. So does the vedalia beetle and, aside from a brief outbreak of scale that (as we shall see) was caused by man's mindless intervention, it has been able to deal with the pest ever since.

In contrast to later enormously expensive, not nearly so successful control programs waged with chemical insecticides, the suppression of the cottony-cushion scale cost the American people about 1,500 dollars: the total of Albert Koebele's passage to Australia and his expenses there and in New Zealand. His name became famous around the world; in Germany, biological control was referred to as the Koebele method. Citrus growers in California presented him with a gold watch and, elated by their good fortune, added a pair of diamond earrings for Mrs. Koebele.

2

LEARNING FROM THE PAST
Time and the Citrus Whitefly

THE WHITEFLY IS NOT A TRUE FLY. A MEMBER OF THE ORDER HO-
moptera, and a relative of scale insects, the morphological resemblance between the adults of the two groups is
detectable only by a specialist. Adult whiteflies, most of
whom are two or three millimeters long, are winged creatures. The white waxy substance they secrete on their backs
and wings (and which clings like powder to the leaves they
inhabit) gives the family its name. Like other members of
their order, they possess sucking mouthparts. Entomologists, in fact, sometimes refer to them because of their traits
as the tropical equivalents of the aphids.

The citrus whitefly (*Dialeurodes citri*) attacks several species of plants, but as its name implies, it specializes in citrus trees. So tiny is it upon hatching from the egg that an unobservant grower may not be aware that his trees are already under attack. In this stage the larva, if it can be seen at all, resembles a miniature louse or a mite, with its three pairs of stubby legs. It wanders briefly over the leaf where it hatched, finally settling down in a congenial place. This is its first and last trip as an immature insect. It sinks its thread-like mouthparts into the leaf, extracting the juices, and as it grows, it sheds its exoskeleton as well as its legs. Like most other insect larvae, it is simply an eating machine.

After the third molt, the citrus whitefly pupates. In this stage it does somewhat resemble a scale insect—a thin, nearly transparent organism plastered to its leaf. As it approaches maturity, the pupa grows more opaque, with two purple spots, which are to be the eyes of the adult, appearing at its forward end. The body thickens and reaches a length of about one-sixteenth of an inch. Upon hatching, the pupa case splits down the back and the adult whitefly emerges.

It is now in the "fly" stage of its life cycle. Its body is a light orange-yellow. Two sets of pure white wings give the adult true mobility and the female may fly some distance to another tree and begin to lay her eggs on the underside of the leaves. She may lay well over 100 eggs. An entomologist who has observed a number of females ovipositing on a single leaf described its surface as "presenting an appearance as though sprinkled with grains of dust." In Florida, the citrus whitefly usually passes through three generations—in spring, summer, and fall—each year.

As larvae, these insects undoubtedly rob a tree of some of its vitality, but the citrus whitefly's most severe impact stems from another characteristic. Like aphids and scale insects, the adult whitefly secretes honeydew. When honeydew falls on the upper surface of a leaf, a sooty mold flourishes and begins to spread like a dark stain. Dust and other fine debris gather in the sticky coating, restricting photosynthesis. Falling onto the fruit, the honeydew and the mold it nourishes disfigure the skin, retard the ripening of the fruit, and sometimes cut off its growth completely.

Unchecked, whiteflies reduce the clear green and gold of an orange tree to shabby ruin. Blackened by sooty mold and sucked dry by tiny mouths, the leaves curl into the tortured state that growers call "rat-tailing" and drop off in the next wind. Serving as a heat sink, the sooty mold produces broad necrotic splotches that render even ripened fruit unfit for market.

The citrus whitefly became, in the early years of the twentieth century, Florida's counterpart of the cottony-cushion scale. By 1910, it was the most serious pest in the state's orange groves and growers said that wherever it appeared in large numbers it reduced the crop from one-fifth to one-half. This whitefly also invaded California, but as yet without serious economic consequences. Because attempts to control the insect with various sprays and fumigants had failed, the time was obviously ripe for biological control.

Matters had not gone smoothly in this new science following its first great triumph. The initial enthusiasm was soon tempered by the realization that biological control was not merely a process of turning a boxful of predatory insects loose in a field or an orchard. Happy surprises, such as the introduction of the vedalia beetle, did not happen every day. In most cases, a great deal of preparation was needed for the successful importation of an effective natural enemy, including studies on the life histories of both the insect pest and its predators and parasites. A good deal of luck was needed, too. The technology of the early twentieth century—including transportation to and from remote areas, as well as the instruments and facilities needed to study and rear tiny insects—was not up to all the demands placed on it by biological control.

Moreover, a clash of personalities helped check the growth of the new science. C. V. Riley, as Chief Entomologist in the federal government, seemed to resent the leading role Californians were taking in the search for natural enemies. When Californians raised money to send Albert Koebele back to Australia to search for parasites on various scale insects, they asked Riley to put up an additional sum to extend his trip. Riley refused, but pressure from leading citizens in California forced the Secretary of Agriculture to override his decision.

"Riley became increasingly bitter toward the California situation and in September 1893 he recalled his agents from California," writes R. L. Doutt. "Coquillett returned to Washington, but Koebele had apparently seen the change coming and had corresponded with the Hawaiian government about employment. When Riley recalled him, he immediately resigned and went to work for Hawaii at a tremendous salary increase ($250 per month and expenses)."

The antipathy between the federal administrators and the Californians did not abate in 1894, when Riley was succeeded as Chief of the Bureau of Entomology by his assistant, L. O. Howard. Howard believed that the success of the vedalia beetle had put stars in the eyes of California's agricultural authorities, causing them to depend wholly on natural enemies to the exclusion of all other insect control measures. Doutt and other scientists have shown that this was not the case. In fact, the federal government did not enact a national insect quarantine program until 1905, almost one-quarter of a century after California's quarantine law. According to Doutt, Howard then threatened to wreck California's biological control program by using quarantine regulations to keep the state from importing beneficial insects. Years later, Howard was still flailing his old foes, and in his *History of Applied Entomology,* published in 1930, he got in a final thrust. "There has been a tendency for many years for persons with strange beliefs to migrate to California, largely on account of the climate," he wrote, "and southern California today is known as the home of all the heterodoxies."

Florida, however, was the area hardest hit by the citrus whitefly, and the federal government attempted in 1910, to find natural enemies that might be effective in a control program. Congress (perhaps softened now by a fellow feeling for predators and parasites) appropriated 5,000 dollars to send a collector abroad. Howard chose one of his bureau's associate entomologists, Russell Woglum, who had been working in California to develop fumigating techniques for use against scale insects. Woglum's experiences as a foreign collector demonstrated the odds against a successful biological control program at that time.

As in most collecting trips abroad, the first step was to determine which region to explore. No one was quite sure where the citrus whitefly had come from. Woglum believed that citrus trees had probably originated in southeast Asia. When a search of the Bureau's insect collections disclosed specimens of citrus whitefly taken from orange trees in the northwestern Himalayas, he decided to make as thorough a survey as possible for the whitefly's enemies in eastern Asia.

Woglum left New York by steamship on July 31, 1910, bound for the Orient by way of the Mediterranean. A part of his assignment was to stop briefly in Spain, at the invitation of the government, to instruct its leading agriculturalists in the techniques of fumigating citrus trees infested by scale insects. In passing, he surveyed the citrus-growing regions of Spain, Italy, and Sicily for the whitefly. He found no trace of it, nor any evidence that the pest had yet invaded southern Europe. From Naples he sailed directly to Ceylon (now Sri Lanka), where again he found no trace of the citrus whitefly.

"Orange trees do not appear to grow with much vigor on this island, and the fruit produced is inferior in size and quality," Woglum noted. "When picked, the rind is perfectly green, although the flesh may be fully matured. This failure of the fruit to color—a condition noticed in other countries lying near the equator—is doubtless attributable to the excessive dampness of a tropical climate."

Woglum entered India through its southernmost port, Tuticorin. He went on by train to Calcutta, which was the site of the Indian Museum, then the largest institution of its kind in the Far East. A search of its entomological collections helped Woglum to narrow his target area. Although there appeared to be some confusion in nomenclature, he compared specimens, and again the evidence pointed to the Himalayan region as a likely home range of the citrus whitefly. Woglum also acquired as much information as possible about commercial orange growing in India, which he learned was "confined almost exclusively to individual or small patches of trees in yards and native gardens, both on the plains and in the hills or lower elevations of the mountains." Native people in the mountains sometimes cultivated a few orange trees scattered among other trees in the forest.

Woglum discovered that the citrus whitefly specimens in the museum were from a place on the frontier so remote that it would be impractical for him to try to reach it that autumn. He decided instead to visit Saharanpur, which had one of the oldest botanical gardens in northern India. There, at the end of October, he discovered the citrus whitefly.

"The insect at this time had reached the pupal stage,"

he wrote in his report of the trip. "Specimens of the fly could be found on practically all trees examined, but the infestation was so light that the insect was in no way a serious pest. Of the insects infesting the leaves only a small percentage was living. The trees containing the largest number of living insects were noted to be the ones with the densest foliage and those protected by large overshadowing ornamental trees. In no instance was a tree affected by 'sooty mold.' "

Woglum was elated to find a small, reddish-brown ladybird beetle, about one-tenth of an inch long, feeding on the whitefly pupae. Although the beetles were widely scattered, he set about collecting them at once. Spreading large sheets of cloth underneath the trees in early morning before the insects had become active, he and several helpers then went about beating the branches with sticks. Later they collected about 200 ladybird beetles that had fallen onto the sheets.

"About 100 specimens of the insect were placed in a small, specially made wooden box containing two chambers connected by an opening about the size of a 50-cent piece," he reported. "One of these chambers was loosely filled with damp sphagnum moss, the other with dry fiber from a palm tree. Such packing allowed the insects free movement and at the same time reduced possible injury from rough usage to a minimum. The box was so constructed as to allow necessary aeration. This box was forwarded to the American consul-general at Calcutta, who placed the same in the personal charge of the captain of a cargo steamer sailing direct from Calcutta to the United States."

Alas, when entomologists opened the box in Florida, they discovered that all the ladybirds were dead. The same fate overtook another collection Woglum sent home by mail. The only other discovery of note that Woglum made at Saharanpur that fall was a brown fungus that also killed the citrus whitefly. The fungus, which he found nowhere else in

India, looked extremely familiar to him. He sent specimens back to the United States, where an expert identified it as a species native to Florida! "A great deal of exchanging of plants, especially of citrus trees, between the botanical garden at this place and certain nursery men in Florida has been going on for many years," Woglum wrote, "and it seems quite likely that the brown fungus was introduced from Florida into India through these exchanges."

From there, Woglum went by train to Lahore, situated on a broad, hot plain in the Punjab and adjacent to the foothills of the Himalayas. Oranges and limes were abundant in Lahore, but the citrus whitefly, although present, was no more a genuine pest than it had been at Saharanpur.

"During the investigation at Saharanpur a few pupa cases of the citrus whitefly were noticed to differ somewhat in appearance from those of normal shape," Woglum said. "Some of these contained very small holes which were of such a character as could easily have been made by a ladybeetle or some other biting insect. However, when large numbers of these abnormally thickened pupa cases were found at Lahore, and always with a small rounded hole in the exposed surface, it was very apparent that this condition was the result of internal parasitism. Considering the type of the host as well as the character of the opening, one was at once led to infer that the parasite was of a hymenopterous species."

The climate at Lahore is extremely hot and dry in summer, the temperatures climbing as high as 120° F. The winters, however, are comparatively cold, the temperatures often dropping into the middle 30s. Since it was already late November, Woglum found that most active insect life had disappeared. Thus he collected leaves to which the old pupal cases still clung and sent them to the Bureau of Entomology in Washington for further investigation.

Then he took off on a whirlwind survey that was something of an odyssey itself. He ranged across northern India, traveling on horseback in the outer Himalayas (approaching the Khyber Pass at Peshawar) and by small covered boats (on which he could sleep at night) in Assam. He discovered citrus whitefly wherever oranges were grown in the region and, invariably, there was evidence of parasitism. But with insect life practically at a standstill in the winter chill, Woglum decided to search more southerly areas until spring returned to northern India.

Leaving Calcutta on Christmas Eve, 1910, he followed a trail of rumors about large citrus groves growing in the vicinity of Moulmein, in Lower Burma, but found only a few dying trees without whitefly. Boarding a steamer at Rangoon, he sailed for Java via Penang and Singapore. This single-minded collector, for the first time in his report, became almost lyrical amid the lushness of the Javanese countryside and the splendid botanical garden at Buitenzorg, although again there were only scattered citrus plantings and no sign of whitefly.

Upon his arrival in Hong Kong in February, he found a cablegram from L. O. Howard, who had examined the parasitized pupal cases Woglum had sent him from Lahore. Howard had found five dead parasitic wasps, light yellow and about one-half a millimeter long, still attached to the accompanying leaves. It was a new species, which Howard named *Prospaltella lahorensis*. The chief entomologist was eager to have live specimens of this parasite, as well as the ladybird beetle that had failed to survive the shipment to Florida, and ordered Woglum to return to India in the spring. Woglum hoped to visit the orange-growing areas around Canton and Swatow in China, but since there was a delay in securing a passport for the Chinese interior, he decided to visit the Philippine Islands first. He never reached China. In the Philippines he was laid low by a disease, possibly malaria, and spent a month in a hospital there.

In April, although not fully recovered from his illness, Woglum hurried back to India. Lahore, the type locality for the parasite he had discovered, was the natural focus of his search. At first his findings were discouraging because an unnaturally severe winter had reduced the whitefly population. Finally, in the more protected parts of a hedge composed of orange trees in the botanical garden, he found the citrus whitefly and small numbers of the living parasite *Prospaltella*. He was confronted now with the major problem to be overcome by all early collectors: how to get his specimens back alive.

At the beginning, if the trip is to be long, two conditions have to be met: The pest insect needs living foliage and the parasite needs a living host. Woglum calculated that the journey from India to Florida would take five to six weeks. Because the entire life cycle of *Prospaltella* at high temperatures is completed in only three weeks, he would have to provide a living environment for the parasite in which one generation could succeed another.

"As small nursery trees are alone practicable for transportation over great distances," Woglum wrote, "it was at once evident that the success of the mission depended on obtaining young trees well infested with the whitefly. Young orange trees were available in sufficient quantities at the Lahore garden, but all were free from living whiteflies. Young fly-infested trees had not been seen anywhere in that country. The problem thus resolved itself into the artificial infestation of the trees."

With the help of a gardener he had hired as an assistant at the botanical garden, Woglum transplanted a number of young orange trees, from one to four feet tall, into earthenware pots. Trouble developed almost at once. The tender green growth that whitefly prefer was attacked by leaf miners and budworms, and shriveled and died. Woglum then procured a second set of nursery trees, over which he raised shelters of light cloth that would let in the sun and keep out pests until he wanted them. Woglum spent the summer preparing his diverse cargo for shipment—nursing the young trees through the leaf miner season, then removing the cloth shelters to allow adult whiteflies to migrate from the adjacent hedge (from which he had removed the preferred tender foliage) to the potted trees. By the middle of October, the trees were well infested with whiteflies and Woglum was able to see the parasites running about on the leaves.

"The parasite prefers the larval stages of its host, but when necessary will oviposit in the pupae," Woglum wrote. "Parasitized larvae and pupae develop a much greater thickness than healthy ones. They also soon lose their transparency, becoming opaque, and this renders them easy of detection. By the use of a lens the parasitic larvae, which are of a whitish cast, can be seen within the whitefly host. On reaching the pupal stage the parasite becomes very dark, almost black, so that at this time parasitized whiteflies containing pupa cases appear very dark. Having attained maturity the parasite eats a small hole in the dorsum of the host and through this opening emerges into the open air."

Woglum also collected live specimens of the ladybird beetle, by then having been identified as *Cryptognatha flavescens,* which had not survived the trip to Florida a year earlier. He hoped for better luck with this shipment because he had now had a chance to study the life histories of his insects and was aware of their short generation time. This time he would be accompanying his collection, which he transported in miniature greenhouses, called Wardian cases. These cases, invented by the English botanist Nathaniel B.

Ward early in the nineteenth century, had helped to revolutionize botanical collecting and propagation, allowing collectors to ship living specimens from remote areas by even the most primitive transportation.

"The base measurement of the cases was approximately 2½ by 3½ feet while the height varied from 4 to 5 feet," Woglum wrote. "Three large holes were made in both ends of the cases toward the top so as to allow a free exchange of air. These holes were covered with fine gauze to prevent the escape of insects. Two small doors were made in each case to be used when watering the plants. These doors were kept open in good weather during the voyage and especially made fine wire-gauze screen placed in the opening. The glass portion of the case was divided into small sections, six on either side. Thick glass was deep set in the heavy frames so as to reduce to a minimum the possibility of breakage enroute."

Woglum set the earthenware jars holding the young orange trees onto a rack of one-half–inch boards in the Wardian cases, thus allowing for the inevitable jolting about on the trip, as well as for seepage of water. The jars themselves were tightly packed with a mixture of sphagnum moss and palm fibers. Strips of board tacked over the tops of the jars held them in place. When the final screws closing the cases were inserted and tightened, Woglum had five miniature orange groves, each with a flourishing interacting colony of whitefly, parasites, and predators, ready to set out on a journey halfway around the world.

On October 20, 1911, Woglum set out from Lahore. The five Wardian cases of natural enemies (and a sixth filled with native varieties of citrus trees for propagation) were loaded on huge-wheeled oxcarts tended by local men dressed in light turbans and long white gowns. At the rail-road station, the cases were transferred to a train for Bombay, from which Woglum and his collection left by steamship for the United States. It was an arduous trip, across the Indian Ocean and up the airless corridor of the Red Sea. In the enveloping deadness, the smoke curled forward from the ship's funnel, the sky was of silver and the sea of pewter, yet Woglum's moist little universes flourished under glass. Because he had been unable to arrange for a direct passage, the Wardian cases had to be transferred to other ships at Port Said and Naples. Woglum fussed over his charges almost every waking hour. Three days out of New York, the ship encountered a fierce storm and the cases, which had made the trip on deck until then, were moved below.

After a voyage of exactly a month from Bombay, the ship arrived at New York on November 28. Woglum superintended the transfer of the cases to an express train, and on December 2, he bore them triumphantly to the government laboratory in Orlando, Florida. He had returned, mission accomplished, with a living collection of parasites and predators.

It was only during the coming weeks that a sense of disaster came over Woglum and his colleagues at Orlando. Through no fault of his own, he had returned to Florida at the wrong season. Whiteflies were inactive in the winter weather of central Florida and the imported natural enemies refused to feed on the pupal stage. The small collection of parasites quickly expired. Only two of the ladybird beetles remained alive until the frantic entomologists were able to rear whitefly eggs in a heated room. Hope flared briefly when the beetles attacked the eggs voraciously and managed to survive the winter. But, as the two survivors were apparently of the same sex, they eventually died without producing eggs.

The citrus whitefly remained a pest in Florida's orange groves until our own time. It was to be more than half a century before anyone attempted to introduce the parasite *Prospaltella* into the United States again, and by then, as we shall see, it was traveling under a new name.

3

NATURE'S MAZE
The Whitefly and the Aggravating Ant

MORE THAN TWO-THIRDS OF THE SUCCESSFUL ATTEMPTS TO control insect pests with natural enemies have been carried out using hymenopterous parasites—essentially, small, stingless wasps. There are thousands of such species; no one knows how many, but a British entomologist, G. J. Kerrich, estimated that there are 500,000, and new species are being discovered all the time.

The insect order Hymenoptera, which means membrane-winged, includes ants, bees, and wasps. Some kinds of wasps are social insects. The paper wasps, for instance, live in large colonies inside nests built of paper, which they make by chewing up small strips of wood and converting them to pulp with their saliva. Other kinds of wasps are solitary, among them the mud dauber, which plasters its nest of mud and saliva in protected places. These wasps have true stingers. The mud dauber, in fact, uses her stinger to inject a venom and paralyze spiders, or other insects, and brings them back to the nest as fresh food for her young.

Some closely related Hymenoptera, often less than one millimeter long, are the parasitic wasps, and they neither build nests nor use their "stingers" directly as weapons. In these wasps, the stinger functions as an egg-laying organ or ovipositor. Entomologists, however, often use the verb *sting* when they refer to a wasp's practice of plunging its ovipositor into the eggs or other developing stages of the species it has selected to receive its own eggs. Other insect parasites come from such orders as Diptera, the true flies, but as with wasps, their value to biological control lies in the undeviating attention with which they search out a single species or a small number of them. Evolution, as it were, has programmed them to fidelity.

Parasite, derived from the ancient Greek, means literally "feeding beside," or "one who eats at the table of another." The word was first applied, contemptuously, to humans, who ate at another's expense. Only later was it applied to other animals. Note that we bluntly call the victim of a predator its prey; we call the victim of a parasite, euphemistically, its host. A parasite that attacks a single species is host specific.

Entomologists point out that, technically, these parasitic insects are parasitoids. Many common parasites, such as the tapeworms or fleas that attack higher animals (including man), live in or on a host indefinitely, and although they often weaken their victim, they seldom kill it. But the

parasitic insects valuable in biological control are interme-
diate between parasites and predators. The parasitoid is, in a
sense, a specialized predator. Once it has hatched inside its
host, it feeds remorselessly, gaining nourishment as the host
weakens and usually perishes. But, unlike a predator that
feeds on large numbers of prey, the developing parasitoid
lives on a single host.

When it emerges from its "mummified" host, the wasp,
or other parasitic insect, mates almost immediately. During
its adulthood, which may last only a few days, the wasp
usually shuns violence and settles down to subsist on nectar
and honeydew. (The adults of some parasitic species, however, turn briefly into voracious predators, stabbing numerous "hosts" and sucking them dry, thus remaining effective
agents of biological control.) The adult then searches for the
host in which it is programmed to lay its own eggs and the
grim cycle begins again.

Parasitoid, then, is a useful descriptive term, but in
general, these small wasps, flies, and so forth, are identified
in the entomological literature as parasites. Neither name
justly celebrates these dragons of the insect world.

Mike Rose led his visitor into a citrus grove on the University
of California's South Coast Research Station in Orange
County. He is a bear of a man, with an imposing nose
springing from a face that is otherwise nearly engulfed by
straggling hair and a rusty beard. They walked between rows
of mature trees: shiny green leaves and, among them, dull
green lumps that one day soon would blaze into a bounteous
lemon crop. At the end of each row, however, where dirt
roads intersected the grove, the foliage was dark and
lusterless.

"Trucks are supposed to slow down going past the groves, but they hardly ever do," Rose said. "They kick up a lot of dust, and it just adds to the dust that blows in here from all the building going on in this part of the county. There's some whitefly in here—see, here's a little clump of larvae on this leaf. But it's dust and ants that are going to kill us if we're not careful."

Rose, Research Associate in the University's Division of Biological Control, had worked for 15 years on a series of whitefly pests. The citrus whitefly did not, until recently, become the scourge in California that it became in Florida, confining itself on the West Coast mainly to dooryard plantings, but for more than half a century it lurked on the fringes, as it were, while nobody could figure out a way to deal with it. Meanwhile, it was joined in California by the woolly whitefly (*Aleurothrixus flocossus*). This pest, which is probably native to the Americas, crossed the border from Mexico and settled in around San Diego in 1966. By this time, the state and federal agricultural agencies had armed themselves with batteries of modern chemical insecticides. The state intended to eradicate both species of whitefly. It spent millions of dollars on spraying programs before giving up in frustration, as the citrus whitefly held doggedly on and the woolly whitefly dramatically extended its range into the citrus country east of Los Angeles.

Paul DeBach, the tall, scholarly entomologist whose name has become almost synonymous with modern biological control, had already imported natural enemies of the two whiteflies when the chemical crews went to work. In 1968, he had used technologies not available to the ill-starred Russell Woglum. He simply imported Woglum's precious wasp, *Prospaltella lahorensis,* by air mail from India and West Pakistan. Modern taxonomists have placed this parasite in the genus *Encarsia,* but a change in name has not reduced its effectiveness. *Encarsia lahorensis* quickly became established in California, where it is now used to fight new outbreaks of the citrus whitefly. Sent to Florida by DeBach and Rose, it is a valuable check on the pest there.

"The sad part of the story," DeBach has written, "is that these natural enemies could have been reimported to Florida and doubtless established many years ago, thus, probably preventing considerable economic loss and reducing or eliminating the necessity for chemical sprays for the whitefly."

Turning to the woolly whitefly (which the orange industry and the state had estimated would cost the growers between 15 and 20 million dollars a year in chemical insecticides), DeBach and Rose traveled to South America to find this insect's natural enemies on its native ground.

"We imported parasites of the woolly whitefly from Latin America," Mike Rose said. "They became established, and the woolly whitefly hasn't cost the growers a cent in treatments. We now have both the woolly and citrus whitefly under complete biological control in California. You will never get complete eradication with either chemicals or natural enemies, but when you find a solution with biological control it is permanent."

Of primary concern to both Rose and DeBach in recent years has been the latest in the series of whitefly invasions, this time by the bayberry whitefly (*Parabemisia myricae*).

"The bayberry whitefly was found here in California for the first time in 1978," Rose said. "It really took off, from San Diego all the way up into Ventura County, and in California it threatened to be the most immediately destructive of the whiteflies. The reason it spread so fast is that males are very rare in this species—there are mostly females, which are winged and parthenogenic.

They are able to produce young from eggs that haven't been fertilized by males. They began feeding on lemons in this area, and on a large nearby ranch they also attacked avocados. In 1979 there was a damaging infestation in this grove we're in right now. But there have been no chemical insecticides used here in more than six years, and we established the newly introduced parasites of the bayberry whitefly here immediately. Just look at these leaves! Except out there near the road where dust and ants have interfered with the work of natural enemies, you will find very few whitefly."

As Rose led the way to his laboratory in a sprawling, one-story building beyond the glasshouses, he told of how he and DeBach had plotted their campaign against the bayberry whitefly.

"The whitefly spread so quickly that it was obvious chemicals weren't going to contain it," he said. "The California State Department of Food and Agriculture asked us to start a biological control program against this insect in 1979. It does a lot of damage because both the adult females and the flat, plate-like larvae are feeding all the time. They congregate on the tender new growth of a tree. On lemon trees this new growth is bright red, like the red bayberry leaves we discovered it feeds on in Japan. Under high density the whitefly will kill the growing tips and stop the fruit from developing. When they're crowded, the females may even lay their eggs on the fruit, leaving it mottled and downgrading it for market."

The researchers at first found no effective natural enemies attacking the bayberry whitefly in California. It is native to Asia, and a search of the literature indicated that it had invaded large mulberry plantations in Japan earlier in this century. (Like lemon trees, which are pruned to promote large fruit and to make it easier to harvest the crop, mulberry trees send out constant new growth because they are cut back severely to provide a continuous supply of tender greenery for silkworms.) There was only one record of a natural enemy, a species of *Encarsia*.

"We've had good relations with Japanese entomologists for a long time so I flew there in 1979," Rose recalled. "I discovered several other parasites had turned up in recent years. One of them was a wasp in the genus *Eretmocerus* that was found around Kochi on the island of Shikoku. We brought several species back, and *Eretmocerus* has turned out to be very effective."

Rose and DeBach returned to Japan in 1981. A lingering controversy among entomologists has been whether to import more than one natural enemy for the biological control of a pest insect. Some experts argue that several parasites competing for the same host would tend to nullify each other's efforts and, ultimately, be a waste of resources. DeBach was in the forefront of those who advocated bringing in the widest possible selection of genetic material, from different regions and climates, and determining in the field which species, or strain, was most effective in the new environment. He and Rose, therefore, explored the area around Nagoya on Honshu because it was known for extreme highs and lows in temperature and humidity. Their object was to collect as hardy a gene stock as might be available among *Eretmocerus* and to obtain other species of parasites.

"You never really know in advance how a species will adapt to a new environment," Rose said. "You can make certain generalizations about genera that you are familiar with. Species in *Eretmocerus,* for instance, are easy to recognize and are often effective natural enemies against whitefly. But species in the genus *Encarsia* are really tricky. Some species are very difficult to identify and you begin to wonder if you are really looking at *Encarsia* after all. There are species in that genus that, under certain reproductive circumstances, behave as hyperparasites—they parasitize other parasites, and you certainly don't want that. Those species must be excluded from our collections.

"So, following a long day of collecting in Japan, perhaps 12 to 14 hours, we would sit for hours and hours back in the hotel room, isolating the pupae of parasites we had found on little patches of leaves; and then we would put each one in a tiny glass vial with a touch of honey so it would have food when it emerged from the pupa. We stoppered the vials with sterile cotton and paper. The vials had to go into little rearing units—Petri dishes filled with a salt and water slurry, sealed with a semipermeable membrane to maintain a high humidity so that the immature wasps did not desiccate. Finally, we hand-carried all of them back to the United States."

Upon their arrival at the Quarantine Insectary at the University of California, Riverside, Rose and his colleagues carefully examined all the developing parasites on a round-the-clock schedule, looking them over for hyperparasites, mating them, and attempting to unravel the details of their life cycles. By the end of 1983, Rose had supervised the release of more than 3 million parasites (*Eretmocerus* and *Encarsia*) in California citrus groves.

Now, in his laboratory, he invited his visitor to look through a microscope at a living *Eretmocerus* he had deftly prepared on a slide for viewing.

"You have a very steady hand," his visitor noted.

"When you work with lively creatures that are less than a millimeter long you develop a steady hand," Mike Rose said. "Those large antennal clubs—they look something like boat oars—are characteristic of the genus. [In fact, the word *Eretmocerus* is put together from the Greek words for "oar" and "horn."] They have fine sensilla, or sense organs, all over the antennae and when the female is searching for a host she holds them high and picks up her cues. What she is looking for is a bayberry whitefly larvae in its second or third stage of development. When she finds one she presses her abdomen down on the leaf and deposits an egg *under* the larva. Her own larva hatches from the egg, uses its jaws to cut an entry hole in the underside of the whitefly larva, and enters its body. It gradually consumes the host, leaving only the cuticle or outer skin, which we call a mummy. The parasite pupates right there inside the whitefly mummy. It develops into an adult, cuts an exit hole, and flies away to find hosts for her own eggs.

"What has happened then is that we have converted this living tissue from a whitefly, which would be out there as a pest on citrus, into a parasite, which is out there hunting down other whitefly. We are in the process of seeing an exponential change in the composition of the biomass of the community."

In the best biological control programs, there is a further step beyond the importation, rearing, and release of natural enemies. That step is evaluation, and taking it distinguishes the scientist from the propagandist. It is not enough for the biological control worker to know that he is out there destroying insect pests. He must convince growers, and even other entomologists, that the system works. Proponents of chemical insecticides are quick to disparage the effectiveness of natural enemies, sometimes arguing that a dramatic decline in a pest population can be attributed to weather or some mysterious natural cycle. Biological control workers have devised methods of establishing the effectiveness of a program with mathematical certitude. Mike Rose pointed to what looked like a jumble of cloth bags on a table in the laboratory.

"We call them exclusion cages," he explained. "Paul DeBach and his mentor, Harry Smith, devised them back in the 1940s to fit over individual branches on a tree for comparison studies. We make them ourselves out of fine parachute cloth. We put them out in pairs on trees in the same area of a grove, even matching them up with the same age structure of the leaves and branches on the tree. One of the sleeves is a true cage, covering a group of whitefly from the egg stage on so that no parasites can reach them. The other sleeve in the pair is open at the ends so that parasites can come and go naturally.

"Otherwise, everything is identical inside the two cages. The whitefly can develop, move around, and feed normally, and climatic conditions are the same. But if parasites are working effectively, you will see a tremendous difference. This method tells us, for one generation of whitefly, what the effect of parasitism is—the rate of parasitism, the survival rate of the hosts, and so forth. We have found that on some trees the survival rate of the whitefly is 100 times greater when they are protected from natural enemies."

Time and again in southern California, Mike Rose has seen the various species of whitefly come under the control of imported parasites. Recently, an *Eretmocerus* species never before found in California has come to light as an effective natural enemy of the bayberry whitefly, apparently having either adapted to the whitefly from another host, or arrived somehow as an immigrant. By every scientific test, the parasites are successful in their one-on-one relationships with their hosts. But Rose is aware that nature is never so simple, that outside forces are always at work tilting the balance in such relationships, and that in biological control there is an element that is cause for concern.

The Argentine ant (*Iridomyrmex humilis*) is an accessory before and after the fact, a collaborator with various scale insects, aphids, and whitefly in making life miserable for growers of everything from corn and sugar cane to oranges and lemons. In the complexity of their lives and the company they keep, Argentine ants could help turn the groves of Arcady itself into a bucolic slum. Natives of South America, these ants were first detected in New Orleans in the late nineteenth century. In fact, for a time they were called "New Orleans ants," to the great indignation of that city's residents. They spread quickly by means of waterways and railroads through much of the South and in 1907 were first collected in California.

Argentine ants introduce a new destructive element in citrus groves that any of several species of whitefly have infested. Whitefly manufacture a honeydew, particularly rich in amino acids and basic sugars, that is irresistible to Argentine ants. These exotic pests, living in colonies with multiple queens, are extremely prolific. The workers, oval-headed insects about two and one-half millimeters long and colored a deep metallic brown, fan out from the new colonies. They quickly discover the source of the sweet, nourishing substance coating the citrus leaves and appropriate "flocks" of whitefly for themselves. They protect them as zealously as any good shepherd tends his sheep. Entomologists believe that the ants may actually trigger the secretion of honeydew in whitefly larvae by biting or stroking the sensitive organs that release it.

The workers gather the sweet stuff and take it back to the nest as food for their young. "The abdomen is capable of considerable distention," writes an observer of an Argentine ant colony that fed on the secretions of various sucking insects, "and when the worker is fully engorged with syrup or other liquid its chitinous plates are forced apart, rendering the connecting membranes distinctly visible. The writer has often noticed workers returning from their attendance on plant lice with abdomens so distended that they looked like little drops of silvery liquid." The colony, fueled by this rich food, grows at a dramatic pace and infests one citrus tree after another.

Argentine ants, although not feeding on the trees directly, compound the grower's problems many times over. Their aggressiveness, and sometimes simply their incessant busyness, frighten away the predators and parasites searching for whitefly. By gathering the excess honeydew, ants save the whitefly from their own productive zeal as well; the latter often produce so much honeydew that a considerable number of them actually drown in a sea of it, and the diligent ants thus remove a potential check on their increase.

Finally, by a curious ramification of the ants' activity, colonies of other pests gain a foothold on the tree. The ants, following chemical cues, wear a clearly defined path from their nests among a tree's roots along the trunk and branches to the haunts of the whitefly on the tender outer twigs and leaves. Entomologists have noticed dense colonies of red scale and other pests springing up alongside the ant trails because potential parasites and predators have been kept away from those parts of the tree.

"A grower I know put in several hundred young lemon trees and banded the trunks with wax paper to protect them from rodents," Mike Rose said. "Well, the Argentine ants just moved in and made their nests in those wax paper tubes. They were harvesting honeydew and, of course, interfering with the natural enemies on the trees. Pretty soon all kinds of pests were erupting. That farmer ought to take down the wax paper and spread something like Tanglefoot on the trunks to keep the ants off the trees."

Entomologists have been working on that problem with some urgency for many years. Poisoned pellets, traps, and other devices have not been completely satisfactory. Mike Rose is probably the only worker in biological control who has traveled to foreign countries in an attempt to find effective natural enemies for the Argentine ant. He spent long hours in the field in South America, observing the ant and collecting potential natural enemies of various pests, sorting and preparing them for shipment. From Argentina he made 13 shipments of natural enemies to California, each shipment requiring payment to three different officials to make sure that his insects were put aboard the proper planes before they dried up and died. In Paraguay, he was arrested for clipping shoots from trees, unaware that they were growing in a national shrine.

"The worst time I had though was in Argentina during a government crisis at the end of Eva Peron's regime," he said. "I was out in the countryside, down on my hands and knees with a small shovel, digging in the tunnels of Argentine ants to see if I could find anything that might be preying on them. I heard a noise behind me, and when I looked up there were two big army officers pointing their pistols at me, and right behind them six wide-eyed, fuzzy-cheeked soldiers holding automatic weapons. I knew enough to keep my mouth shut and just show them my papers. They thought that I was planting some kind of bombs. They even started to take apart my microscope, but somehow I got them to let me do it instead. They took me off to jail. I stayed there for several hours until my interpreter spoke to them and got me released."

Although the problem of ants on citrus trees has not been completely solved, Rose believes that growers can, by persistence and attention to detail, keep them from dominating their groves. Having grown up on a farm in Ohio, where little reliance was put on chemical insecticides, Rose still prefers to fight nature with nature.

"The *grower* has to become a good *farmer,*" he insisted. "If he pays attention and recognizes problems before they get a head start, he can forestall most of them. Too many growers who are influenced by pesticide salesmen panic at the first sign of insects and turn to insecticides. When a grower sprays citrus, he is jeopardizing the results of 90 years of biological control research in this part of the country."

4

THE DREADED GYPSIES

Attacking the Forest Killers

THE EGGS OF THE GYPSY MOTH (*LYMANTRIA DISPAR*) HATCH IN late spring, when a tiny, blackish caterpillar, bristling with long hairs, chews its way to freedom. Spinning a long silken thread, it suspends itself from a tree and is often carried long distances by the wind. (An observer reported that one caterpillar after hatching spun a thread 69⅓ feet long.)

The young caterpillar grows rapidly, feeding at night on a great variety of deciduous leaves, especially oak, birch, poplar, and fruit trees, and in times of shortage resorting to pine and some other conifers. By day, it crawls off to hide in bark crevices or in leaf litter. This is the gypsy moth's marauding stage, when it defoliates to one degree or another large tracts of woodland. By early June, the caterpillar has reached a length of about two inches. To the objective observer (of whom there seems to be a great scarcity), it is a rather handsome beast, big-headed, its back dotted with prominent blue and red spots toward the rear, and with tufts of pale yellow hairs sprouting from its sides. After it finishes feeding, the caterpillar pupates in an out-of-the-way place, to emerge in several weeks as a moth.

As adults, these insects are prime examples of sexual dimorphism—the development of very different physical characteristics in males and females. The male moth's ruddy-brown color, according to popular etymology, was likened in England to a gypsy's complexion and gave the moth its vernacular name. On the European continent, it has carried a number of common descriptive names, including the German *Grosskopfspinner* (great-headed spinner) and the French *le bombyx disparate* (the dissimilar silkworm, a reference to those striking differences between the sexes that also give an element to its scientific name).

And the female is dissimilar, a much larger moth than her mate, white with black zigzag lines on its broad wings. She is, however, generally flightless, apparently unable to lift her large body, packed with hundreds of eggs, off the ground. (In the Far East the female *can* fly.) She simply crawls a few inches from her cocoon and attracts a mate by emitting a powerful and irresistible scent.

After mating, the female lays her 500 or more eggs in a cluster, sometimes more than 1,000 if she happens to be on the fringe of a vigorous population that is rapidly expanding its range. Each cluster, one inch or more long, is left in a protected place, perhaps the underside of a stone, a branch, or a log. The female moth fashions a covering of brownish-orange fuzz for the cluster from the hairs and scales of her abdomen and then dies.

The variation between sexes is repeated among the many strains of this species. For a long time, the gypsy moth occasionally served as an alternative to the fruitfly and other common insects in genetic experiments, since it is able to maintain itself in widely different habitats, extending across Europe, North Africa, and Asia to Siberia, China, and Japan.

The gypsy moth is one of the tussock moths, lumped by entomologists into a group called the macro-moths that includes such families as those of the sphinx moths, tent caterpillars, royal moths, and the silkworm moths. And that, as far as the inhabitants of many parts of the United States are concerned, was the origin of the problem.

In 1869, a French astronomer and naturalist, Leopold Trouvelot, who was living in Medford, Massachusetts, imported from Europe the eggs of a number of moths to investigate the possibility of developing a silk-spinning caterpillar with more resistance to disease than the domesticated silkworm, pehaps through hybridization. Among the eggs were those of the gypsy moth. When the eggs hatched, he put caterpillars on shrubs in his backyard where they could feed naturally while confined by nets he had bound over the foliage. The disaster began as it usually does in horror movies—with heavy rain, thunder, and lightning. The storm that struck Medford that early summer evening tore the nets from Trouvelot's shrubs. The alien creatures escaped, spreading inexorably throughout the town.

For about 10 years, gypsy moths fought their lonely battle with the New World's environment, and little was seen of them. Then, suddenly, they began to swarm in the neighborhood of Trouvelot's former home. The eyewitness reports, a century old now, that those long forgotten citizens of Medford left with the authorities about their municipal plague would serve nicely in a newspaper report describing a modern outbreak.

"We moved to Medford in 1882," a woman recalled. "The caterpillars were over everything in our yard and stripped all our fruit trees, taking the apple trees first and then the pears. There was a beautiful maple on the street in front of the next house, and all the leaves were eaten . . . they got from the ground upon the house and blackened the front of it. . . . We destroyed a great many caterpillars by burning, but their numbers did not seem to be lessened in the least."

By 1889, they had spread all over town. "My sister cried out one day, 'They are marching up the street,' " another woman said. "I went to the front door, and sure enough, the street was black with them, coming across from my neighbor's . . . and heading straight for our yard. They had stripped her trees, but our trees at that time were only partially eaten."

Homeowners gathered the caterpillars in baskets and burned them. Pedestrians felt them fall from trees onto their hats or down their collars; some residents carried umbrellas even on fine days. Women complained bitterly as they trod dozens underfoot and returned home with their long skirts stained by squashed caterpillars. A man reported the caterpillars "were so thick on the trees that they were stuck together like cold macaroni." As fruit and shade trees died and property owners complained, the state finally intervened. E. H. Forbush, the noted New England ornithologist, was named "Field Director in Charge of the Work of Destroying the Gypsy Moth." Forbush led a vigorous campaign that, some later entomologists believed, might have finished off the troublesome insect because it was still confined to a reasonably small area.

Manpower was cheap then, and Forbush put it to work. Crews searched out egg masses and painted them with creosote. They cut and burned small, heavily infested woodlots. They sprayed the caterpillars with arsenate of lead. One of the most effective measures in parks and around homes was to tie a burlap bag, its flap hanging down, around the trunk of a large tree; at daybreak, when the caterpillars moved down the trunk in search of hiding places, they crawled under the burlap bag where they could be easily collected and burned.

As early as the 1890s, a few New Englanders, having heard of California's remarkable success with predators and parasites, advocated a search for the gypsy moth's natural enemies abroad. But the authorities in Massachusetts seemed to believe by then that the war had been won and prematurely abandoned the many-pronged attack on the invader. Only 20 or 30 of the 10,000 leaf-eating caterpillars in North American forests are serious pests. But the gypsy moth, unchecked by the enemies and diseases that in the Old World held it in reasonable control, advanced (often borne by the wind) through the northeastern woodlands, where decades of man's abuse may have left the trees especially vulnerable to attack. Three-fourths of the forests in Massachusetts and Rhode Island have burned at least once in the past century or so. Such fires destroy the diversity of woodland species, and favor the survival of the thick-barked, fire-resistant oak, which in many areas became the dominant tree. A monoculture, as we know from agriculture, is particularly susceptible to invasion by pests. And of all our forest trees, none seems to be more attractive to the gypsy moth than the oak.

As the gypsy moth spread through the forests of eastern

Massachusetts early in this century, foresters and entomologists supplemented their early methods of control with more sophisticated weapons, chiefly parasites. They imported certain wasps and tachinid flies known to parasitize gypsy moths in their native habitats in Europe and Japan. They also released a large green beetle, *Calosoma*, that proved an efficient, host-specific predator within a limited range. There was an added and extremely important dividend for American forests in those early importations. The only reliable method of shipping parasites in those days of long voyages by steamer was to pack the parasitized hosts themselves with fresh foliage in Wardian cases or the like. Sometime before World War I, a virus that is deadly to gypsy moths entered the United States in one or more of those shipments and became established in the wild. During gypsy moth outbreaks in subsequent years, this contagious disease spread like wildfire through the population and effectively decimated the pests.

Although the gypsy moth spread to several neighboring states from Massachusetts, it seldom caused serious problems. Occasionally, as in the 1920s, the population surged to peak numbers and then collapsed from starvation, disease, parasites and predators, and climatic conditions. Through the 1930s and the 1940s, the population seemed to stabilize itself, fitting less disruptively into the forest community. Although there was occasional defoliation, the gypsy moth plagues of the past century were becoming ancient history.

Then after World War II, DDT became available for general use. The USDA, giddy with this latest manifestation of man's technical ingenuity, decided to use DDT to blow the gypsy moth off the face of the earth, or at least that portion of it within its jurisdiction. And so, wherever the "eradication" specialists detected this pest, the spray planes appeared and laid down their chemical barrage. DDT

Rachel Carson first warned of toxic showers.

worked fantastically well, killing gypsy moths far more efficiently than anything else, before or since. But always,
within a year or so, there was another outbreak. Congress
appropriated money and the spray planes roared in once
more over the tree tops, pursuing the USDA's goal of complete eradication.

"They sprayed truck gardens and dairy farms, fish
ponds and salt marshes," Rachel Carson wrote in *Silent
Spring* as she noted the spray planes' scattershot approach.
"They sprayed the quarter-acre lots of suburbia, drenching a
housewife making a desperate effort to cover her garden before the roaring planes reached her, and showering insecticide over children at play and commuters at railway
stations. . . . Automobiles were spotted with the oily mixture;
flowers and shrubs were ruined. Birds, fish, crabs, and useful insects were killed."

Ironically, it was during the heyday of DDT that the
gypsy moth threw off its natural shackles to begin the dramatic extension of its range that has continued into the current decade. Was this simply a coincidence? Reece I. Sailer,
who was active in gypsy moth studies as a biological control
specialist with the USDA for many years, spoke of that insect's "suddenly violent fluctuations" during the 1950s. "I
can't help but feel that we were seeing evidence of human
interference with the natural balance that had stabilized the
gypsy moth's population in the 1930s," Sailer told a government conference some years later. "If you spray over a large
area you annihilate all the parasites. We went from a policy
of coexistence to one of annihilation, and we have not been
successful."

The outcry against DDT's side effects reached a cre-
scendo when public health authorities warned that excessive amounts of the long-lasting chemical were appearing
in milk. Reluctantly, the USDA and its cooperating state for-

estry departments gave up that discredited chemical for gypsy moth control and switched to another insecticide, Sevin, which breaks down more quickly than DDT and, thus, is not recycled through food chains.

The fatal flaw in this single-minded chemical approach remained. Like DDT, Sevin tends to perpetuate the problem so that spraying must be continued year after year, with only short-term relief. Enough gypsy moths survive each chemical treatment to produce the eggs that ensure a bumper crop of pests the next summer. Moreover, insecticides kill enough caterpillars so that there is no competition for food and very little opportunity for disease to spread; under these conditions, the females lay egg masses containing unusually large numbers (perhaps 600 or 700) of healthy eggs. But the moth's natural enemies recover more slowly (for reasons to be discussed in Chapter 18) from a massive spraying program. This is especially true of parasites because, even when a few survive, the alternate hosts they may depend on for part of the year are wiped out and they, too, decline. The use of Sevin has other unfortunate side effects. This chemical is extremely toxic to bees, destroying entire domesticated colonies belonging to honey producers as well as wild colonies on which many plants depend for pollination. Although Sevin apparently is not extremely toxic to birds, its use often coincides with the birds' nesting season, thus killing off the variety of insects they require to feed their young.

By the early 1970s, it was obvious that the use of insecticides to eradicate the gypsy moth was at a dead end, but few alternatives seemed to be available. "The research and management effort languished during the DDT era," admitted a high official of the U.S. Forest Service. "Now we do not have adequate management techniques for the present resurgence." The USDA and various state agencies turned once more to science. They began intensive research into alternate controls, such as the pathogen *Bacillus thuringiensis*. Other scientists worked to synthesize the pheromone released by female gypsy moths to attract males; the resulting product, called Disparlure, is now an important weapon set out by scientists to confuse males or to lure them to traps where they can be sterilized or destroyed.

After many years of neglect, the USDA revived its program for importing parasites. Both those natural enemies and the scientists who worked with them were held in low regard among the USDA's decision-makers during the time the agency promoted chemicals as the final solution for pest insects. Now men like Reece Sailer (who almost alone among USDA scientists cooperated actively with Rachel Carson when she was writing her book) could speak out and have at least a fair chance of finding their counsel accepted by administrators. They urged that the search for Old World parasites be pursued more forcefully.

Roger Fuester is a large man with a quick sense of humor and an unbounded enthusiasm for tracking down and putting to work the natural enemies of pest insects. For six years he was assigned to the USDA's European Parasite Laboratory outside Paris, searching for parasites of the gypsy moth. He found both the work and the environment delightful, although he is almost apologetic that his European exploration produced "no vedalia beetle" for the gypsy moth.

"Yes, I'd like to go back some day, but our son was just finishing kindergarten, so that seemed like a good time to come back and let him start school over here," Fuester said. "USDA thinks six years is enough, too—they don't want you to become expatriated! Anyway, looking for gypsy moths in Europe is an entirely different game. Europe doesn't have the tree mortality we have, or the acreage that's hit by the moths. After all, over there we may be talking about defoliation of 14 or 15 acres, not 40,000 acres.

"There are gypsy moths in France, but they're seldom a serious problem. It's a funny thing, though—the climate's suitable, but there's absolutely no problem in England. We put out a number of traps baited with the synthetic sex attractant Disparlure on the French side of the English Channel, and we picked up quite a few males. If a male is anywhere in the area we'll get it with that stuff, but we didn't come up with a single one in all our traps on the British side of the channel. Do you know what I think? England is full of people who collect butterflies and moths. This is an attractive moth which has never been very common in England anyway, and I think that as soon as they appear over there they go right into someone's collection. Whatever the reason, I wish they'd disappear like that over here!"

Fuester was an ardent collector of gypsy moths himself while he was in Europe, chiefly for the prizes they might contain. It was his assignment to hunt down the small, isolated populations that had no economic significance in the hope of discovering parasites that kept the gypsy moth at such low levels. During the summer of 1972, he and two colleagues surveyed France and West Germany, and two summers later they traveled through much of Austria, mapping the locations of small populations.

"We set little traps on tree trunks, about six feet above the ground, in various forests and woodlands," he said. "Each trap had a cotton wick that we impregnated with Disparlure. We would leave them out anywhere from four to fifteen days and then check the contents. We found that the gypsy moth exists throughout most of Germany. In Austria it seems to be confined mainly to the eastern part of the country. We caught two in the Tyrol, which were the first gypsy moths ever recorded there because as a rule they're not found in high altitudes. The two we captured were in low country along the Inn River."

Fuester's surveys brought a windfall of parasites that

emerged from the gypsy moths he found all over Europe. He collected and shipped to the United States 14 species of parasites from Poland alone. Perhaps the most interesting of his discoveries was of a nematode, *Hexamernis albicans*, which infected and killed gypsy moths in various places throughout West Germany and Austria. At first he could not keep the nematodes (which are roundworms) from desiccating after they had emerged from the host larvae. Eventually he and his colleagues solved the problem by covering the bottom of their container with wet paper towels. Then they packed the nematodes in large Petri dishes, spreading coarse vermiculite between wet paper towels, and shipped them to the United States. The first release of an imported beneficial nematode in North America was made during an outbreak of gypsy moths in New Jersey in 1974.

"Every little bit helps," Fuester said. "I classify natural enemies of the gypsy moth in three categories. One is an organism that is host specific; the second is an organism that attacks the gypsy moth, but will also accept other hosts; and the third is an organism that has a very broad range of hosts. We have very few of the second category in this country, compared to the Old World, and I think that's why they get more consistent control over there. That second category can't control the gypsy moth by themselves, but they may take 20 to 30 percent of them—and that's a lot!"

Fuester believes that the intensive exploration for parasites of the gypsy moth in Europe and Asia during the past decade has nearly exhausted the possibilities.

"I think the two best areas that may be left for exploration are mainland China and the southern part of the Soviet Union, west of China and east of the Urals," he said. "We've gone about as far as we can in looking for magic bullets."

5

THE DREADED GYPSIES

Concocting a Lethal Cocktail

ASIDE FROM INSECTICIDES, THE SINGLE MOST DISRUPTIVE ELE-
ment in a specific biological control program may be the hyperparasite—a parasite of a parasite.

Entomologists go to great lengths to screen out hyperparasites from their imported colonies of natural enemies.
(All hyperparasites are wasps and often look much like the
primary parasite they attack.) Just as parasites are extremely effective destroyers of plant-eating insects, "hypers"
may take an enormous toll of these primary parasites. An entomologist in Delaware, searching fields for an established

parasite of an alfalfa pest, found that one-third of the parasites' cocoons had been invaded by hyperparasites. Because this was early in the season and those hyperparasites produce two generations a year, the percentage of beneficial insects they destroyed must have been much higher.

Thanks to the vigilance of entomologists in the quarantine laboratories through which beneficial insects enter the United States today, the carnage is not nearly as severe as it might otherwise be. Most hyperparasites found in North America are native species. They tend to have wide host ranges, however, and a hyper that has been parasitizing native beneficial insects may quickly switch over to an exotic parasite recently imported for a biological control program.

A dramatic example of hyperparasitism occurs in association with the pea aphid (*Acyrthosiphon pisum*). A wasp in the genus *Aphidius* parasitizes the aphid. After an *Aphidius* larva hatches inside the aphid, another wasp, this one in the genus *Alloxysta,* comes along and probes the aphid's integument with her sensitive ovipositor. If *Alloxysta* detects the presence of the *Aphidius* larva inside the aphid, she deposits her egg *through the body walls of both insects* inside the larva. Thus while the *Aphidius* larva is busily consuming the aphid, the *Alloxysta* larva is at its equally lethal task inside the *Aphidius* larva.

But the story is not over yet. There is a third wasp in the genus *Asaphes* that acts as a tertiary parasite, or a parasite on a hyperparasite. It drills a hole in the mummified covering of the now-defunct aphid, as well as the skin of the stricken *Aphidius,* and deposits an egg on the surface of *Alloxysta.* The *Asaphes* egg hatches into a larva that feeds on *Alloxysta* through its body wall and eventually kills it.

Some hyperparasites seem to act merely on faith. They attack an insect that generally serves as a host for a given species of parasite, although it has not yet been parasitized.

If the primary parasite eventually arrives to parasitize the host, the larva of the hyperparasite is already inside, awaiting its victim. If the parasite never arrives, the hyperparasite perishes. Either way, it's a deadly game.

Just as nature has provided many valuable crops with a complement of pests that attack their various stages, human ingenuity has devised a battery of natural enemies to undermine the gypsy moth's dominance in susceptible forests. In the absence of a "magic bullet"—a single lethal instrument as was the vedalia beetle against the cottony-cushion scale—entomologists in a dozen states are at work observing long-established parasites, evaluating the assets and deficiencies of newly discovered ones, formulating stronger dosages of pathogens, and weighing the potentials of sterilants and pheromones. One of the most active of these dragon hunters is Ronald M. Weseloh of the Connecticut Agricultural Experiment Station.

Weseloh, a slender man with dark hair just beginning to turn gray at the temples, has been on the staff of the experiment station since 1970. There, in one of the red brick buildings set among greenhouses and neatly clipped lawns on the edge of New Haven, he maintains an experimental colony of gypsy moths on an artificial diet concocted of wheat germ, casein, and salts; he also rears a limited number of parasites. The results of his studies illuminate the difficulties that even a well-coordinated control program faces in dealing with such a wide-ranging insect as the gypsy moth.

"I was born in California but I grew up in Utah, and that's where I became interested in insects," Weseloh said. "I began collecting them when I was in the sixth or seventh grade. When I attended Brigham Young University I majored in zoology. My old interest in insects was still there, and so my faculty adviser suggested that I investigate biological control. That's why I went to the University of California at Riverside. Yes, I took a class there with Paul DeBach. I did my Ph.D. in host selection behavior—the elements that influence a parasite to attack a particular host and the cues by which it finds the host. I guess this is a theme that has run through all my later work."

Weseloh has specialized in the gypsy moth and its natural enemies since coming to Connecticut. The Connecticut Agricultural Experiment Station is entirely independent of the state university and is funded separately by the state legislature. Its gypsy moth control program dates to the turn of the century, when the moths were blowing into the state from Massachusetts and the only defense consisted of setting up barrier zones and destroying egg masses. The hopes of the station's entomologists rose with the introduction of parasites from Europe and Asia, beginning in 1905 and continuing with the eradication attempts using DDT after World War II; they subsequently languished with the realization that there was no single panacea. Since Connecticut banned all aerial spraying of chemical insecticides in the 1970s, the experiment station has become deeply involved in biological control. Weseloh's projects (some of them in cooperation with other scientists) have included the evaluation of new parasites, the role of hyperparasites, and the interaction of parasites and pathogens.

"Parasitic wasps have always received the most attention in biological control, and rightly so, but other insect orders have supplied a number of successful parasites, too," Weseloh said. "Tachinid flies, which are true flies, often prove to be excellent parasites and the U.S. Forest Service has used one species with a good deal of success against the gypsy moth right here in Connecticut. I've studied another tachinid fly, *Compsilura concinnata,* which was imported around the time of World War I as a parasite on both the gypsy moth and the brown-tail moth. A large number of parasites have been brought in to fight the gypsy moth since the early part of the century, and in some cases all we know about them is whether they became established or not. A lot more needs to be known about how successful each of the established parasites is in controlling gypsy moths."

Ron Weseloh drew on the research of others, as well as on his own observations, to evaluate *Compsilura* as a gypsy moth parasite. This fly is convenient to rear in the laboratory because it mates easily and accepts a variety of hosts besides the gypsy moth; it also produces a number of individuals from each host. Once he had a flourishing colony, Weseloh framed his experiments as any scientist does to answer a series of specific questions: Is *Compsilura* likely to come in contact with gypsy moths in the wild? What stages of the moth is it likely to attack? Does it parasitize an acceptable percentage of gypsy moths? He sought to answer the questions, in part, through field observations, in part, through a series of complex experiments, and from the answers he obtained he was able to make a decision about the tachinid fly's value to a control program.

Weseloh designed one of his experiments to determine where *Compsilura* carried out its search for a host. He knew from the experiments of other workers that tachinid flies often use plant odors as a cue, being attracted to trees or other plants on which their hosts generally feed. He selected oaks and beaches at Sleeping Giant State Park near the experiment station, in an area where gypsy moths were abundant and therefore could be expected to attract parasitic insects. To each tree he nailed boards at regular intervals to

A robber fly is one of many gypsy moth predators.

form a ladder and then tethered gypsy moth larvae with black sewing thread at various levels on the trunk and foliage. After three days, he brought the larvae back to his laboratory and dissected them to see if they had been parasitized.

At the same time, Weseloh carried out experiments in the laboratory, using an ingenious contraption to determine the time of day *Compsilura* was most active in seeking a host. He confined mated females in a cage with honey and water. Through slits in the cage he fed a strip of brown paper to which he tethered gypsy moth larvae at intervals with thread and cellophane tape. The strip rested on rollers and was pulled slowly through the cage so that any one larva remained exposed to the parasites for about 80 minutes. Periodically, Weseloh removed the larvae from the strip and reared them individually in plastic cups to see whether parasites eventually emerged from them. By noting the time of day when each larva had been exposed, he was able to calculate the period when the parasites were most active. In still another experiment, he exposed different instars, or developmental stages, of the larvae to *Compsilura* to determine the size of host it preferred.

The results of his experiments gave Weseloh a picture of *Compsilura*'s behavior in the wild. The parasite attacked larvae mainly on the lower leaves of trees and was active mainly by day, behavior that coincided best in place and time with the early instars of gypsy moth larvae. (The later instars feed higher in the canopy and mainly at night, hiding themselves in bark crevices and leaf litter during the day.) Although *Compsilura* attacks all instars of gypsy moths if they were available, Weseloh found that for some reason its young did not grow as fast in the smaller larvae, even though their chances of survival were the same.

"One of the big problems here is that this tachinid appears to have low population densities in the spring when the young gypsy moth larvae are present," Weseloh said. "So, if an outbreak is beginning to build, the proportion of these parasites to gypsy moths in the area is very low. That and the fact that *Compsilura* has a number of other host species means it parasitizes only about 1 percent of the gypsy moth population. So it probably isn't worthwhile to release any more of these parasites."

In 1979, entomologists released two new parasites of the gypsy moth in Connecticut. The USDA had imported these two small wasps, *Coccygomimus disparis* and *Brachymeria lasus,* from Japan, where they had been found in gypsy moth pupae. Since 1905, ten imported parasites of this pest had established themselves in North America, although only one imported since 1933 had been successful.

"We wanted to follow these two parasites from the start to see how they adapted to conditions here and, if they didn't, what the problem was," Weseloh said. "We don't have mass-rearing facilities at the station, so they were reared for us by the New Jersey Department of Agriculture. They were chosen for release in Connecticut because their range in Japan indicated they might possibly be able to overwinter here."

Again, the results were disappointing. Like most other gypsy moth parasites, the young of these species emerged from host pupae and quickly dispersed so that little could be discovered about their long-term behavior in the wild. However, Weseloh and his colleagues made use of cages containing hosts in the release plots to gather some preliminary information. For instance, *Brachymeria* tended to attack hosts confined in sunny areas and, later on, failed to survive the winter in cages left outdoors. *Coccygomimus* showed a higher tolerance for the New England winter, but preferred other hosts over the gypsy moth.

"Our preliminary studies suggest that neither of these parasites will be entirely satisfactory for gypsy moth control in Connecticut," Weseloh said. "They don't seem to have become established here, but we haven't given up on them yet, and we'll continue to search for them around the release sites."

A far more successful natural enemy, one which Weseloh has studied for many years, is *Cotesia* (formerly *Apanteles*) *melanoscelus*, a larval parasite of the gypsy moth that was brought to New England from Europe in 1911 and has since spread into most infested areas of North America. Because it has been studied for so long and, in recent years, so intensively, *Cotesia* serves to illustrate the assets and drawbacks of an exotic parasite that has never achieved a dramatic success, yet forms a valuable part of the gypsy moth control program. Sportswriters would describe *Cotesia* as "a journeyman performer."

Ron Weseloh used *Cotesia* at first to pursue his special interest in host selection behavior. In a series of experiments, he demonstrated the importance of the silk produced by gypsy moths in helping *Cotesia* track down its host. The parasite showed little interest in the silk spun by other species of moths, but responded with searching behavior—examining the area intently by movements of its antennae—even when it was exposed to leaves chewed earlier by gypsy moths. (When Weseloh's assistant, working with forceps and a dissecting microscope, removed traces of silk from the chewed leaves and presented only the leaves to *Cotesia*, the parasite responded feebly.) Like other entomologists, Weseloh believes that knowledge of this kind might be put to good use in control programs, applying concentrated extracts of the silk to areas where they hope to attract additional parasites.

Although *Cotesia* is an effective hunter, its life cycle

makes it attractive to hyperparasites. *Cotesia* goes through two generations a year in New England, the adults of the first generation emerging in spring from overwintering cocoons and parasitizing early instars of the gypsy moth. The second generation emerges from hosts in June, attacks larvae that are still for the most part in their early instars, and then diapauses within solitary cocoons from mid-July until the following spring. Although the overwintering cocoons are hidden within crevices in tree trunks and under bark flaps, they remain extremely vulnerable to both predators and hyperparasites. Mice, voles, and shrews appear to eat a large number of cocoons, as do birds when the cocoons are spun high in the trees. But the toll taken by hyperparasites is staggering. Two early observers found 35 species of hyperparasites in *Cotesia* cocoons and speculated that as many as 99 percent of the cocoons may be destroyed during *Cotesia*'s long defenseless stage. Weseloh, setting out *Cotesia* cocoons by gluing them to tree trunks in various situations, found that 50 percent of them were parasitized in a single week; others had been ripped apart by unknown predators.

Cotesia's place in Connecticut's gypsy moth control program is made even shakier by its inability to adapt readily to the life cycle of its host. Because of its tremendous losses over winter, *Cotesia* emerges in the spring in low numbers just at the time when the early instars of the gypsy moth are abundant. Weseloh's laboratory studies show that *Cotesia* prefers these smaller larvae to larger ones. *Cotesia* does extremely well under the circumstances, parasitizing a reasonable percentage of the pests. In warm temperatures, the young parasites develop rapidly within their hosts, but in the cooler temperatures characteristic of a New England spring, their growth slows considerably. By the time the first generation of parasite adults appears in Connecticut, many gypsy moths have developed to the fourth instar, when their aggressive defensive movements and long hairs discourage *Cotesia* from parasitizing them.

Weseloh points out that the strain of *Cotesia* established in New England was imported from Sicily. Its growth within the host is still apparently tied to Mediterranean temperatures and is therefore better synchronized with Sicilian gypsy moths than with those in New England. He still hopes that collectors will find and ship to New England a strain from a harsher climate.

The search for a better adapted strain of *Cotesia* might be said to parallel the work of other researchers in recent years to develop more effective formulations of the bacterial insecticide *Bacillus thuringiensis* (*B.t.*). Encouraging results have been obtained with *B.t.* in recent years, and large numbers of gypsy moths have been killed during outbreaks, saving much of the area's foliage. Yet studies by Weseloh and others at the Connecticut Agricultural Experiment Station, the U.S. Forest Service, and the Pennsylvania Department of Environmental Protection also suggest that *B.t.* and *Cotesia* may work well in tandem.

"A substantial proportion of the gypsy moth larvae may survive the early application of *B.t.* but get what we call a sublethal dose," Weseloh said. "The toxic crystals associated with *B.t.* paralyze the gut of these caterpillars and prevent them from feeding, at least temporarily. This action slows down the growth of the caterpillars just enough so that they don't reach the fourth instar as early as they normally do. We have found significantly higher rates of parasitism in areas that have been treated with *B.t.* There is a definite synergistic effect there: *B.t.* kills enough of the gypsy moths to provide us with a more favorable ratio between parasites and hosts. In addition, *B.t.* has slowed down the development of the caterpillars so that *Cotesia* is not overwhelmed by their numbers."

Biological control specialists, like the parasites and predators they work with, are opportunistic creatures. Ron Weseloh believes that such findings as the synergism between *B.t.* and *Cotesia* offer bright new possibilities for the campaign against gypsy moths.

"This study shows dramatically how important the size of the host is for efficient parasitism," he said. "We think it is possible now to manipulate the *development* of the host to cause substantially greater parasitism. We may be able to do this inexpensively in the future with nontoxic feeding deterrents and perhaps even with low doses of *B.t.* We would combine this approach by mass-rearing *Cotesia* and releasing the parasites into the areas that have been treated with *B.t.* There's still a great deal of research to be done, but we will pursue some of these possibilities."

6

DEADLY CARGO
The Battle Against the Cereal Leaf Beetle

NO ONE IS CERTAIN HOW THE CEREAL LEAF BEETLE (*OULEMA melanopus*) entered the United States. A quarantine against agricultural pests is maintained at all of our ports of entry, and suspect cargoes are either fumigated or turned back. Quarantine officers, in fact, intercepted this pest on at least six occasions between 1957 and 1962. But once during that time it slipped through the net.

There is some evidence that it was present in Michigan in 1959. In 1962, it was positively identified in southern Michigan's Berrien County, and within a year or two agricultural specialists considered it a major pest wherever it appeared. Like almost everything else of prominence or notoriety in this country, it began to be referred to by those in the know simply by its initials—CLB.

The adult CLB is, like so many members of the order Coleoptera, a rather attractive creature. (There are people who collect and mount dried specimens of especially handsome beetles, as other people keep for home exhibition a collection of semi-precious stones.) It is about three-sixteenths of an inch long and hard-shelled, and its head and wing covers are a shiny blue-black. Its legs and the segment of its thorax just behind its head are reddish-orange. The larvae, which are generally more destructive in grain fields than the adults, are appropriately unhandsome. A trifle longer than their parent, they have fleshy, half-moon-shaped bodies somewhat resembling the larvae of the Colorado potato beetle.

Adult cereal leaf beetles spend the winter hibernating in field stubble, fence rows, and thickets. As the days grow warmer in early spring, the beetles emerge from their hiding places and seek out the tender young grains and grasses in nearby fields. There they begin to feed and mate, the females laying their eggs on the leaves.

When the larvae hatch, they too begin feeding on the leaves. Later they drop to the ground, where they burrow into the soil and pupate. Adults of the new generation appear in late June and July and feed intermittently on the grain leaves until the cool days of autumn drive them into hibernation. In spring the cycle begins again.

It was a warm, hazy June morning in the rolling farm country near La Porte, Indiana. The climbing sun had spread a glimmering pallor across the east, while thunderclouds began to mass somberly half a sky away. A dozen or more men and women walked into an eight-acre field that was hedged by poplars and woody thickets and planted to small grains, each one carrying clippers and a large cylindrical plastic container. They set to work among the young oats at once, reaping their curious harvest, clipping handfuls of the emerald leaves and setting them loosely upright in the containers.

"We'll have to work fast," said Tom Burger as he studied the sky. "They'll be lucky if they can get any of those little planes up once it starts to storm."

Burger slapped vigorously at a mosquito that was adding its irritating note to the heavy air, but it was not the only insect on his mind that morning. A stocky, energetic man with close-cropped brown hair, and a native of Michigan, he had started out in junior college studying game biology, but discovered his lifelong specialty when his faculty adviser coaxed him into registering for a class in entomology that was in danger of being dropped by the college for lack of students. He has gone on to become one of the USDA's specialists in biological control.

Like the other federal and state agriculture specialists in this Indiana grain field, Burger had come to collect oat leaves studded with the eggs and larvae of the cereal leaf beetle. The insects were feeding voraciously on the leaves, chewing their way through the field as they have done in other fields on three continents and sometimes causing the crops to be plowed under. Yet these oat leaves and their pestiferous cargo were going to be distributed throughout Indiana and its neighboring states. This day had been planned for a long time.

"The oats through here are full of the larvae," Burger pointed out to several officials who had come from the

USDA's headquarters in Washington to observe this "field day." "See, they've got that frosted look to them."

The beetles' eggs, clear yellow cylinders laid singly on the upper surfaces of the leaves and darkening with age, stood out against the bright green oats, although they were less than one-sixteenth of an inch long. The larvae also looked dark until Burger rubbed away the brown fecal matter in which they encase themselves (like spittlebugs in their froth) to ward off desiccation; then their plump bodies were seen to be yellowish, with black heads and legs. A few of the metallic blue adults were visible, too. As Burger had said, the leaves of the oat plants looked "frosted" where the larvae (which seldom eat completely through the leaves as the adults do) had chewed out long strips between the veins, leaving only the white lower epidermis.

When the workers had filled their containers with leaves, they covered them with moist cloth, sealed them with tape, and stacked them in waiting vans to be driven to the airport. According to plan, planes provided by the state would fly the containers to airports around Indiana, where they would be met by county agents who would distribute them to farmers in their districts. The farmers, in turn, would carry the oat leaves into their fields and stand them next to their own growing plants, which would sustain the larvae once the cut leaves wilted. If all went well, by early evening supplies of the destructive beetles' eggs and larvae would be spread throughout every grain-growing region in Indiana.

But the beetles and not the farmers were the planned victims of this plot. The eggs and larvae shipped out on oat leaves that morning carried deadly freight. Inside many of them was a tiny wasp that even then was growing to maturity by consuming the tissues of its host. The progeny of the cereal leaf beetles were transporting the seeds of their own destruction. Upon emerging, the scattered wasps would multiply, and other generations would carry on the USDA's crusade against the pest in years to come.

Thunderstorms soon became general all over Indiana. Some of the planes could not carry out their missions that morning, and the USDA vans delivered as many of the precious containers as they could to the county agents. But Burger and his colleagues looked for a better day later on in the week, for the scientists who wage war on pest insects with biological control are accustomed to meeting and surmounting thorny problems every step of the way.

"The object of man's game with nature is not to win, but to keep on playing," a wise scientist has said. The war on the cereal leaf beetle tidily illustrates the complexities that scientists must deal with both before and after they have decided to turn to biological techniques. At first, the entomologists found themselves trapped on "the treadmill of chemical control." Then they turned to nature for help. The two methods are not always mutually exclusive, but the final emphasis on an alternative to chemicals, as shown here, provides our best hope that we can keep on playing the game.

Like so many of our serious insect pests, the CLB is not native to North America. It inhabits most of Europe and extends across the Soviet Union all the way to Siberia. For a long time it has been known as a pest on small grains such as wheat, oats, and barley. Since invading North America, however, it has also attacked corn. In this age of worldwide food shortages, any threat to grain is cause for immediate concern.

As a rule, insect pests from abroad are first detected in coastal regions, close to seaports or international airline terminals, where they have reached our shores hidden away in plants, seeds, grain, lumber, and other products. Scientists call the CLB a "jet-age insect" because it is the first serious insect pest to come directly to the Midwest from the Old World without having been discovered first near the coast. It may have arrived in the Great Lakes on a ship sailing through the St. Lawrence Seaway, or even in cargo carried by jet to Chicago or Detroit.

As these destructive insects multiplied and spread through the Midwest during the early 1960s, the agricultural community took arms against them. Spray planes swept over grain fields, showering them with Malathion, an organic phosphate insecticide. The USDA established a quarantine (prohibiting growers from transporting untreated hay, fodder, small grains, corn, sod, and harvesting machinery from an infested plot) designed to confine the beetles to the area in which they had been discovered, but neither the chemicals nor the quarantine stopped them.

"These are flying insects," Tom Burger said. "They'd get up in the air, and the prevailing winds, which are southeast in this region, carried them as far as 30 miles a year—over the quarantine and out of reach of the Malathion. We sent up researchers in small planes, and they stuck nets out the windows and trapped CLBs at 1,500 feet. We eventually dropped both the quarantine and the federal spray program, though some individual farmers keep using chemicals when the beetles build up in their fields to high levels."

With the wind as their ally, and encountering no effective natural enemies, the cereal leaf beetles fanned out from Michigan into Indiana, Illinois, Wisconsin, Ohio, Missouri, Kentucky, Tennessee, West Virginia, Virginia, Maryland, Pennsylvania, New Jersey, New York, and Massachusetts.

"They haven't spread very much west of the Mississippi," Burger said. "The winds haven't been right, I guess, and the beetles prefer smaller fields with lots of cover nearby, rather than the wide open spaces of the West. They haven't gone into the Deep South yet, either."

But the USDA's biological control specialists detected a ray of hope in the cereal leaf beetle's background. Although the beetle has caused occasional problems in Europe, particularly in the Balkans and the southwestern Soviet Union, Europeans do not consider this insect a serious pest. Apparently it has made its peace with the environment, erupting in periodic outbreaks, then subsiding into insignificance when natural controls go to work on its inflated population. The challenge to the USDA was to find those natural controls.

For many years, the USDA maintained a European Parasite Laboratory near Paris. This facility was the center of the agency's effort to collect and rear parasitic insects that might help to control exotic pests that had established themselves in the United States in the absence of their natural enemies. But many variables affect a search for a suitable beneficial insect. A parasitic species may control a pest insect in Italy, but may not be able to function in the ranges of temperature and humidity peculiar to Michigan, West Virginia, or New Jersey. It must have a reproductive capacity that can respond to a rapid increase in its host's numbers. It must also have the ability to find its host quickly; an insect that has trouble locating a host at some distance from where it matures will be of little use as a biological control.

The USDA scientists and those cooperating with them in many countries on both sides of the Iron Curtain began an intensive search for CLB larvae in their known habitats, picking them by hand from the leaves of small grains and other plants. These collections were made in a variety of climatic regions so that the researchers would have available a wide range of genetic material in each of the parasite species. A European Baedeker might be compiled simply from the places CLB larvae were found: Saragossa, Rome, Cahors, Seville, Villefranche, Kiev, Lidkoping, Gstaad, Ybbs, Frederikshavn. The collectors shipped the CLB larvae to the laboratory near Paris, where technicians dissected them or kept them under observation to see if parasites finally emerged.

The beetles, their larvae, or their eggs were hosts to at least eight parasites. Not all were selected for further study in the United States. For instance, a tachinid fly found parasitizing CLB larvae at only two locations, Rome and Ybbs (Austria), was already known to attack the larvae of other species and, in fact, had been introduced into the United States in 1939 in an effort to control the asparagus beetle. It had not been able to establish itself then, and scientists believed it would be ineffective against the CLB. Researchers, analyzing the prospects of another candidate, found in their search of the literature that the parasite was not by any means host specific, since it attacked many species of moths and leaf miners.

Four species of parasites showed much promise, all of them gnat-sized wasps that lay their eggs in the eggs or larvae of the cereal leaf beetle. When the young hatch, they devour their unwitting host. At this point, the researchers had to make a decision. Should they concentrate on one likely candidate, or choose several for further testing in the field? Some scientists leaned toward the introduction of a single species on the theory that competition among four or five parasites might reduce the effectiveness of the most efficient species.

Yet there was more to be said on the other side. Time was important because the cereal leaf beetle was spreading rapidly throughout the Midwest. If for some reason, the chosen parasite failed to adapt itself to life in the United States, several years would have been wasted. Moreover, four of five species were apt to provide more complete control throughout the CLB's extensive range, whereas locally they would furnish more stability.

Ultimately, four species of parasitic hymenoptera were

Parasites attack egg and larva of the cereal leaf beetle.

chosen to carry the USDA banner in the war against the cereal leaf beetle. No initials were bestowed on these tiny and practically unknown soldiers, nor even any chummy popular cognomens; they linger in the obscurity of lengthy scientific names. Three of the wasps, from the genera *Diaparsis, Lemophagous,* and *Tetrastichus,* parasitize CLB larvae. The fourth, from the genus *Anaphes,* parasitizes the eggs.

Before the final candidates were chosen, researchers at the European Parasite Laboratory needed to understand more clearly how some of the little wasps might fare once they were released in the United States. For instance, would two or more of the larval parasites lay their eggs in the same host, and if so, what would be the fate of the offspring? While dissecting a CLB larva collected in the field, a researcher might find eggs of two parasitic species. When the young parasites hatched, they fought, and one generally killed or wounded the members of the other species. But it was concluded that such encounters are probably uncommon because the three individual species tend to lay their eggs at different times.

A more serious drawback might have been the ability of some victims of parasites to "encapsulate" the eggs laid in their bodies by their enemies. R. J. Dysart, who dissected a great many CLB larvae in Europe, found that *Diaparsis* eggs are more frequently encapsulated by their host than the other two larval parasites under investigation, indicating that one species of parasite somehow incites a more vigorous reaction in the host than the other species.

"In dissections, eggs and larvae of all three parasite species occasionally were found encapsulated," Dysart wrote of the CLB's defense. "The encapsulation is formed by blood cells of the host larva which congregate about the surface of the parasite egg or larva. Additional haemocytes add themselves to the mass which gradually acquires a translucent, cobbled appearance . . . encapsulation of individual parasites varied in degree from partial to complete. Partially encapsulated parasite larvae often were found still alive though unable to move freely. When parasite larvae were found completely encapsulated, they usually were dead. Completely encapsulated parasite eggs are presumed to be incapable of hatching."

During the late 1960s and early 1970s, nearly one-quarter of a million parasites of the four species were shipped from Europe to the United States. In most cases, they were sent for further screening at the USDA's Introduced Beneficial Insects Laboratory at Moorestown, New Jersey. The adults, captured after emerging from the parasitized CLB larvae kept in cages outside Paris, were packed with honey and water in small containers and shipped air freight from Orly Airport in Paris. Within 10 hours they were picked up by USDA scientists at Philadelphia International Airport and taken to Moorestown. Because of the speed of transfer, a great proportion of the wasps arrived healthy and vigorous. Researchers, releasing them in cages where CLBs were held, watched the female parasites mate and then immediately find, and lay eggs in, their traditional host.

Other wasps arrived in Moorestown as larvae in diapause (the "resting" stage of their life cycles) within CLB pupal cells. Researchers carefully examined these shipments to make certain the parasites themselves had not fallen prey to other species of parasites.

"We don't want any parasites on *our* parasites," one of the scientists said.

But Moorestown was a long distance from the heart of the cereal leaf beetle's range. The USDA officials wisely decided to build a laboratory in enemy territory, where the parasites could be reared and released conveniently near the threatened grain fields. Niles, a small city in southern Michigan near the Indiana border, was chosen as the site of one of the most complex and successful biological control programs ever devised.

7

THE LABORATORY
The Collapse of the Cereal Leaf Beetle

SCIENTISTS ADD EVERY YEAR TO THE "BELIEVE-IT-OR-NOT" DOS-sier they are compiling on the parasitoids, or parasitic insects. Five insect orders have produced these amazingly specialized creatures: Hymenoptera (wasps), Coleoptera (beetles), Diptera (true flies), Lepidoptera (moths), and Strepsiptera ("twisted-wing" insects, which are closely related to the beetles). To date, entomologists have found the Hymenoptera most adaptable, and thus most useful, in biological control, with the Diptera (especially the tachinid flies) a distant runner-up.

Most of the better known parasitic insects are endo-parasites, developing inside the bodies of their hosts. Ecto-parasites develop outside the host, inserting their mouthparts into the host's body and drawing out sustenance, as mosquitoes filch our blood. Some species of parasites deposit only a single egg in or on a host; others, called gregarious parasites, may deposit several or even dozens of eggs.

Within these broad categories of behavior are some bizarre variations. For many years, entomologists were baffled in their attempts to rear certain species of *Coccophagus,* a genus of chalcid wasps and an important parasite of the mealybug. Stanley E. Flanders, of the University of California, Riverside, finally deciphered the riddle of their life cycle in the 1930s. All males of these "deviant" species are hyper-parasites—that is, they feed on other parasites they find occupying the same host. If a female *Coccophagus* is already present, the male does not hatch until his "sister" has consumed the host, whereupon he emerges to feed upon her. Thus, females, belonging to the dominant gender in their species, occasionally sacrifice their lives to make certain that enough males survive. Entomologists politely term this sort of behavior adelphoparasitism, from the Greek word for brother, *adelphos.*

The four parasitic wasps selected for the cereal leaf beetle campaign exhibited no such extravagant tastes, differing from each other chiefly in their preference for a particular stage of their host's development. Three of the four species were parasites of the CLB's larvae. The female wasp mounts its victim, thrusts her ovipositor through the fecal matter in which the host has enveloped itself, and plants one or more eggs in its body. The fourth species "stings" the CLB's eggs. Each pursues its host with relentless attention. A researcher observed one of these wasps "occasionally stinging the hosts through the plant leaf from the opposite surface." Once the young hatch, they begin life with a similar drive, so that in cases where several parasites inhabit a host, the competition becomes so fierce that only one is apt to survive.

But even parasites behaving in such a straightforward manner can cause complex problems in the rearing laboratory.

The Niles Laboratory opened in 1966: a low, homely building of concrete and glass, holding serviceable cubicles for desks, temperature-controlled rooms for insect cages, and brightly lighted rooms for trays of growing oats and other host plants. The USDA eventually dubbed it the Cereal Leaf Beetle Parasite Rearing Laboratory, an unwieldy though provocative name that became one of the first obstacles the staff had to overcome.

"The news that we were raising parasites and spreading them around the countryside scared the hell out of some people," said Tom Burger, who was hired by the USDA to join the laboratory's staff, and who later became its director. "Of course, those people were thinking of tapeworms or cattle ticks or something like that. It's a good thing they didn't know they were *wasps* besides!"

As soon as the Niles facility opened, the quarantine staff at Moorestown began shipping parasites there. The adults, having emerged from their hosts in the Moorestown cages, were collected by gentle vacuum suction into glass tubes (each tube with a small-diameter "collection tip" at one end and a cheesecloth cover at the other). They were then blown directly into cylindrical, one-pint cardboard food cartons that served as shipping containers. These cartons, as for all such shipments of beneficial insects, were elaborately prepared. After dipping them in paraffin to increase rigidity, the handlers cut a circular hole in the bottom of each one and glued over it a strip of plastic screen. Inside, they affixed a wet cellulose sponge to provide moisture for the traveling wasps, and scattered wood excelsior around for resting places. Finally, before securing the tight-fitting lid, they spread honey droplets on its inner surface.

Not all the parasites were shipped to Niles. Although the laboratory there became the headquarters of the cereal leaf beetle program, scientists at Purdue and Michigan State universities played an active role in both the research and the day-to-day chores of rearing the parasites and getting them into the fields. The entire program was carried out on a shoestring, with the USDA giving it none of the funds or publicity it lavished on so many of its large-scale chemical "insect eradication" programs launched during the quarter of a century following World War II. The program's success was due almost entirely to the kind of cooperation, ingenuity, and drudgery seldom seen in federal programs.

"There were a number of federal and state agencies involved in the program, but all the people were compatible," remarked Reece Sailer. "The cereal leaf beetle program is a model of interagency cooperation."

Very little was known about the parasites dropped in the staff's lap at Niles. Hundreds of thousands, and eventually millions of wasps would be needed to mount a successful campaign against the rapidly spreading beetle. There were already indications from the USDA laboratories in Europe and Moorestown that several of the parasites would be difficult to rear under laboratory conditions. Modern entomology was geared for killing, not nuturing, insects. Because the USDA had, for the most part, turned its back on biological control since the advent of DDT, there was a scarcity of experienced personnel in that discipline. Many of the staff members at Niles (the total staff amounted to four or five scientists and a dozen technicians) had to be trained in biological control. Then there were problems to be solved about the nurture and care of these wasps.

The staff scientists, at least, had varied backgrounds in entomology. Tom Burger, while studying at Michigan State, had worked for the Entomology Department, setting up and rearing insect cultures for faculty use. He also carried out field trials for chemical companies to determine the effects of their products on insects that attack crops or trees. During the summer, he had worked for an association of high-bush blueberry growers in Michigan, surveying populations of pest insects in their fields.

"I was toying with the idea of getting a job in the chemical industry," Burger recalled. "That's where the big bucks are, you know. But the job USDA offered me in Niles sounded so fascinating that I took it. I knew I would be getting into some exciting work at the very beginning."

To fill the openings for technicians, the staff often hired local men and women who could be trained on the job. Vera Montgomery heard about the new laboratory when it opened in Niles, only a few miles across the state line from her home in Indiana. Because she had always been interested in plants and animals, she was curious about what was going on in the lab, as remote as it seemed from her daily life. She and her husband had raised eight children, several of whom had already left home to go to college. She thought of herself as a housewife with little time to devote to an outside career.

Vera Montgomery had grown up on a farm, where her father taught her a great deal about the natural world and how to raise plants. As a girl, she had wandered around the farm, learning the names of the wild plants and animals and a little about the way they lived. After her marriage, she maintained this interest in wild things, taking her children on walks and telling them what she knew about the creatures they saw. When she could not answer their questions, she went to the public library and took out books on natural history. The lessons stuck, because all of her children pursued careers in medicine and biology.

She and her children often talked about the new laboratory in Niles, wondering what sort of work the entomologists did with the notorious beetle and its parasites. Then one day two of her sons brought home interesting news.

"Guess where the woman across the street got a job?" one of them asked her.

The other son did not even wait for his mother to guess. "At the insect lab," he told her.

A few days later, Vera Montgomery, a shy, dark-haired woman with a soft voice, went to the laboratory and applied for a job. Somewhat to her surprise, for she had no college degree or formal scientific training, she was hired. Her first job was picking cereal leaf beetle eggs off oat leaves.

The parasites shipped to Niles from Moorestown were used as the breeding stock for the CLB program. The laboratory's chief function was to rear enough parasites to inundate the areas of the country infested by the CLB and to make a significant dent in the CLB population. A requisite for rearing any animals, of course, is an adequate supply of food, and in this case the required food was CLB's. The staff had no trouble collecting the leaves of small grain plants infested with the beetles' eggs and larvae in the nearby countryside, and the leaves began to pile up in the laboratory.

Montgomery worked quickly at one of the long tables, picking the small yellow eggs of CLBs off leaves with a dissecting needle. Whenever she had a few minutes, she watched the tiny egg parasite *Anaphes:* released into a glass case containing a supply of CLB eggs, each female visited at least five or six of them and deposited several eggs of her own at every stop.

"I learned to tell the males from the females," she said. "The females have knobs on their antennae, and the males do not. *Anaphes,* you know, spends most of its life cycle in the egg of another species of insect—first as an egg itself, then as a larvae feeding on the beetle's egg, and then as a pupa. When it becomes an adult it makes an opening in the beetle's egg and emerges. I would watch the male *Anaphes* stand around and wait for a female to come from a beetle's egg. Then they would mate and the life cycle would start all over again."

Because she learned quickly, Montgomery was assigned to a new project. Staff scientists had discovered another parasitic wasp, *Trichogramma,* in some of the CLB eggs brought in from grain fields and wondered if it might be a promising candidate for the program.

"I was asked to rear this parasite in the lab for a number of generations," she said. "The hope was that if we just kept supplying it with CLB eggs, after many generations it might become so accustomed to them that this population of *Trichogramma* would attack no other kinds of eggs. I reared them for a long time but it didn't work out. When we switched to the eggs of moths or any other kind of insects, *Trichogramma* went ahead and attacked them. So then we knew that if we wanted parasites to attack only the eggs of the cereal leaf beetle we had to depend on *Anaphes.*"

But there was one positive result of that experiment. Vera Montgomery, who had come to the laboratory in middle age, knew that she could make a career for herself in biological control, and so did her superiors. They gave her time off to take biology courses at the University of Indiana at South Bend. Later, the USDA paid her tuition for entomology courses at Notre Dame.

It became increasingly apparent that the program's success depended on learning through trial and error, as staff members ventured into unexplored territory. They had to overcome one problem after another. Because of its long diapause, when all activity ceased, the CLB itself was the bottleneck in the attempt to raise parasites quickly. There was nothing to do but pack the beetles away in cold storage for many weeks until they became active again. Even so, many of them died in diapause.

Anaphes, on the other hand, could be raised quickly if provided with eggs. It has a brief life span: Within nine days, it develops from an egg into an adult less than one millimeter long. As an adult, it lives only two or three days in the laboratory, just long enough to mate and produce a new generation. The question arose, Would *Anaphes* accept an alternate host that would be faster to rear than the CLB?

Many parasites, such as *Anaphes,* have evolved to be host specific. If only one or a handful of parasites prey on one kind of host, competition is generally restricted and there is always a plentiful supply of food for their offspring. But collectors in Europe had sometimes found *Anaphes* parasitizing a close relative of the cereal leaf beetle. Perhaps this wasp would accept the egg of another relative in this country.

The Niles researchers collected the eggs of several beetles related to the CLB and presented them to *Anaphes.* The parasite showed a clear preference for the three-lined potato beetle (*Lema trilineata trivitatta*), a native of North America with a life cycle of 30 or 40 days, one-fifth the time needed to rear the CLBs. The only trouble was, the potato beetle didn't care much for small grains. It did, however, like jimsonweed which is, ironically, a coarse, rank-smelling, poisonous plant that originated in the tropics, and now is a pest in U.S. soybean fields. Like most "weeds," it is easy to grow, and soon it was thriving in flats under the laboratory's artificial lights.

Anaphes was willing enough, but the potato beetles' eggs presented it with two formidable problems. It had difficulty moving about on the glutinous material with which this beetle coats its eggs, and it sometimes was unable to penetrate the tough chorion, or egg covering. The staff could actually see through a microscope the tiny wasp's ovipositor bending as it tried to sting the resistant egg.

Tom Burger and his colleagues dealt with the first problem themselves. They concocted a solvent of mineral spirits and a detergent that, used under air agitation, simultaneously eased the task of removing the eggs from the leaves of the jimsonweed and washed away the glutinous covering. This solvent worked splendidly, except for a temporary setback brought about inadvertently by an environmental crusade in the early 1970s. Under attack by environmentalists, the company that manufactured the non-biodegradable detergent used in the process suddenly changed its formula. The potato beetles' eggs reacted adversely to the new formula, and many were lost before the staff found an acceptable substitute.

Natural selection went to work and solved the second problem. At first, reproduction by *Anaphes* on the potato beetles' eggs was low because, even when one of the wasps managed to penetrate the tough chorion, its offspring often was unable to chew its way out. *Anaphes* responded by upgrading its vigor.

"There is evidence that natural selection has occurred in *Anaphes* colonies that have been reared on the substitute host," Burger and his colleagues wrote. "A colony placed on the substitute host for the first time increased more than fourfold after seven successive generations. Apparently, weaker and less vigorous parasites that found it difficult to oviposit in eggs of *L. trilineata trivitatta* were selected out of the population in each generation."

Going a step further, the researchers made certain that natural selection remained on their side. "We still went on rearing CLBs for *Anaphes,* because we were afraid that after four or five generations the wasps would get hooked on potato beetles' eggs," Burger said. "So every few generations we put *Anaphes* back on CLB eggs. Sure enough, whenever we gave *Anaphes* a choice, it opted for its traditional host, the CLB."

Meanwhile, the three species of wasps that parasitized CLB larvae proved impossible to rear in the required numbers in the laboratory. They were reluctant to sting the host, except in close confinement, but under crowded conditions many of them died. The scientists at Niles and at Michigan State then developed field insectaries, liberally "seeded" with both small grains and CLBs, where colonies of parasites could proliferate under natural conditions.

"It takes at least three years to get a field insectary started," Burger said, "but the parasites do a terrific job once they start to build up their populations. They keep the CLB population so low that we have to bring in new supplies of CLBs to keep the insectary in operation."

The aim of all this concerted effort was to put parasites in fields where they were most needed. Some growers of small grains discovered densities of five CLBs a square foot in their grain fields, and there were cases where the infestation reached more than two hundred a square foot. The staff, after collecting parasites from the containers with pipettes, transferred them to plastic vials stoppered with wet cotton plugs and took them to the infested fields. There the vials were opened and placed upright next to the plants. The escaping wasps invariably found their prey close at hand. In other cases, as we have already seen, workers from the Niles laboratory and cooperating universities conducted field days on which infested leaves, with their cargoes of parasitized CLB eggs and larvae, were collected and flown to distant sites.

Another important step was evaluating the program. Researchers frequently checked fields where releases had been made to see if the parasites had established themselves. They collected CLB eggs and larvae and took them to Niles, where the staff analyzed them. What proportion of them were parasitized? Which parasites were most effective in the field?

At first, this was an expensive and time-consuming process. Because of their small size and their varied appearance at different stages, the wasps were difficult to identify. Vera Montgomery, working with a young colleague named Penelope DeWitt, made an important contribution at this point. They examined thousands of larval parasites in all stages of growth and published an illustrated paper, in the *Annals of the Entomological Society of America,* that enabled other workers to distinguish among these wasps during their immature stages.

There was no guarantee, of course, that the imported parasites would be successful against the CLB once they were released. They would be operating in an unstable situation, searching for a pest that is migratory and whose host plants are annuals.

"Under Michigan conditions, grain crops are planted in small fields often widely distant," wrote USDA researchers. "Rotation of wheat and oats is a common practice. Thus, host crops of the beetle may not be present in the same location for two successive years. These unstable conditions are not favorable to parasites that must search for their host from field to field during the season, and from one season to the next."

Nevertheless, the release sites and times were chosen so carefully, protecting *Anaphes* and the three larval parasites from insecticide applications and disruptive plowing or disking of fields, that they survived in large numbers. They soon became firmly established in the United States, especially expanding their ranges to the north and east as they followed prevailing winds. (Because the parasites do not buck those winds, they were artificially introduced in greater quantities to the south and west.) By the late 1970s, one or more species had been recovered at nearly 1,300 sites in 15 states and Ontario (to which *Anaphes* spread, although it was not released in Canada). *Anaphes* has destroyed up to seven out of every ten CLB eggs in some fields. One of the larval parasites, *Tetrastichus*, revealed an amazing capacity for migrating and forming dense populations, even in grain fields heavily disturbed by plowing and disking. As one researcher said, "This parasite is just blowing out of the Midwest into the East." It followed the cereal leaf beetle into New England, and in more heavily infested areas, such as Pennsylvania, it was found to parasitize between 90 and 98 percent of the CLB larvae in some fields.

By the end of the 1970s, when the USDA pulled out of the program, it had spent a total of 15 million dollars on the biological control of the cereal leaf beetle. At the height of the infestation, the CLB had been costing growers that much in a single year. Few growers in the Midwest spray for this pest anymore: Agricultural specialists at Purdue and Michigan State keep an eye on local outbreaks and introduce parasites as necessary.

"A few years ago we'd send two collectors out for a few days and they'd come back with 100,000 CLB larvae," Tom Burger said in 1982. "This year we sent out two men and they came back with 135 larvae. Every one of them was parasitized."

8

THE PARASITE HUNTER
AND
THE ICE PLANT SCALE

THE ICE PLANT SCALE WAS RECOGNIZED AS A PEST IN CALIFOR-
nia for several years before entomologists realized that it had
a double identity. It is really two species of insects, *Pulvinariella mesembryanthemi* and *Pulvinaria delottoi*, both apparently originating in southern Africa. They attack,
logically enough, the ice plant, a fleshy perennial in the
genus *Carpobrotus* introduced in quantity to California in
recent times as an ornamental plant and, more importantly,
a groundcover.

The two scales illustrate anew the complexity of mod-

ern insect pest problems as they take on international ramifi-
cations. Even professionals cannot distinguish between
these two species on physical grounds, having to rely on recently acquired knowledge on the timing of their reproduction and development. (In the laboratory, entomologists sort
them out under a microscope, almost solely on the basis of
the differences in the structure of their setal hairs, or bris-
tles.) They both belong to the "soft" scales group, which
does not manufacture a waxy covering during its adult
stage, as do the armored "hard" scales.

Yet, once past that level of similarity, there is some controversy about their systematics. G. De Lotto, an Italian entomologist who has worked in South Africa for many years and
for whom the newly discovered species is named, has placed
them in separate genera, a decision specialists in California
are not happy with. To complicate matters, two other closely
related scales have been described in the Mediterranean region and in Australia, and so there are now four destructive
look-alikes scattered among three genera.

Only *Pulvinariella mesembryanthemi* has been known
to science long enough to have acquired any sort of bio-
graphical record. This scale was originally described in
southern France in 1829. But it is an indication of its pica-
resque past that the first serious description of its biology
was put together many years later in Argentina, a circum-
stance that helps muddy ideas about its land of origin. A
more distinctive profile is emerging from California, where it
has been observed in the company of its sister species.

Both species are parthenogenic, producing offspring
without mating. A few males of *P. mesembryanthemi* are
seen occasionally in warm weather or in greenhouses, but as
researchers have ventured, "males probably represent a rel-
ict natural history character and do not play a role in reproduction." All attempts to mate the males and females of this

species in the laboratory have failed. As for the other species, *Pulvinaria delottoi*, researchers in California have found only one male, who was detected emerging from a laboratory culture.

Mature females of both species produce anywhere from 300 to 1,700 eggs. The emerging first instar nymphs, called "crawlers," measure from one-half to a little more than a millimeter in length. Here the two species diverge in their preferences, the young scales of *P. delottoi* settling down to feed (hence, they are "settlers") on the older stems and leaves of ice plants, whereas those of *P. mesembryanthemi* cluster on new leaves and terminal shoots. Thus, the latter are more frequently observed. This species typically produces two generations a year, but because its life cycle speeds up in warmer climates, it produces three or four generations annually in Southern California. *Pulvinaria delottoi* typically goes through only one generation.

"There's a certain nervousness or stage fright when you begin a search for specimens in a foreign land," said Richard L. Tassan, an entomologist in the Division of Biological Control at the University of California at Berkeley. "The land itself is strange to you, and you begin to think, suppose after coming all this way I don't find anything. It reminded me a little of the way I felt when I was a kid playing football. There'd be that attack of nerves until the game started and the first physical contact was made, and then everything would settle down."

Tassan is a small, wiry man with long hair and glasses. He made his first extensive trip abroad as a collector under uncertain circumstances, looking for a complex of plants, pests, and natural enemies that were very little known or understood in modern entomology. It was a chase in which identity was often an illusion. Certainty had fallen victim to the biological jumble that humans have produced by moving useful or ornamental plants (and inadvertently their faunal entourage) from one continent to another. Tassan's assignment was further complicated by the physical sensation that the world had turned upside down.

Horticulturalists have found a place for the ice plant in private gardens and around condominiums and light industrial developments; the U.S. Army has used it for dune stabilization, as at Fort Ord to bind the dunes behind firing ranges where blowing sand was fouling the mechanisms of movable targets; and the California Department of Transportation has planted it extensively along freeways and interchanges. This all-purpose plant is known in the state by a number of vernacular names, such as beach apple and sea fig, but the people who work with it generally call it ice plant.

"That name is really a misnomer," Tassan said. "The plants we have here are relatives of the original ice plant, *Mesembryanthemum crystallinum,* an annual with small round leaves covered with vesicles, or tiny gland-like dots, that glisten in the sun and make the leaves look as if they are coated with ice. The plants in California don't glisten like that—you can only see the vesicles through a microscope. There's still a lot of confusion about the systematics of these plants, as well as their pests. We have a pretty, yellow-flowered ice plant, *Carpobrotus edulis,* but then there's a whole slew of plants with purple, lavender, and pink flowers of all sizes. Some are hybrids, some may be 'good' species, or the differences in size or color may simply be functions of the conditions under which the plants are growing. Nobody has figured them all out yet."

One group of ice plants in the genus *Carpobrotus* seems to have the center of its natural range in southern Africa, another group in Australia. As plantings became more common in California, scale insects naturally made their appearance, too, in at least one case as early as 1949. Rich Tassan became acquainted with the scale in the early 1970s when he investigated an infestation in two private gardens in Napa, where a number of ice plants were killed. State agricultural officials apparently had controlled the infestation with applications of the insecticide Malathion. But in the mid-70s, the scale erupted in plantings along the state freeways. Caltrans (California Department of Transportation) approached the Division of Biological Control.

"We had good rapport with Caltrans," Tassan said. "We had done a lot of educational work with them about the benefits of biological control, and they had already given us a grant to work on another problem. Besides, they were under a lot of pressure from the public to cut down on spraying."

Caltrans had surveyed its 6,000 acres of ice plants for the scale and had tried insecticides without success. The use of insecticides along freeways is a costly business because traffic lanes must be blocked and other safety precautions taken. Caltrans estimated its insecticide cost at about 75 dollars an acre, with two or three applications needed each year. The annual bill for spraying ice plants amounted to over 50,000 dollars in the Bay Area alone.

The Division of Biological Control maintains a small fund for foreign exploration. In 1978, Tassan was assigned to scout likely regions for the scale's natural enemies, but his destination was by no means clear. The scale had not yet been discovered in Australia, and in fact, it had not been definitely established that there were two scales in California. The first ice plant scale had been discovered in southern France, so that region was on Tassan's list.

"We knew that two parasites of the scale had been collected in South Africa," Tassan said, "and there was one record of a secondary parasite—or hyperparasite—on one of the scale parasites from there. This indicated to us a fairly long relationship between the scale, its natural enemies, and the next trophic—or feeding—level, so that was basically why we selected South Africa for investigation. Also, we had a good contact with their scientists. Some of them had done their graduate work here, and others had cooperated in sending specimens to us. Rhodesia was mentioned to us as another possibility, but I just rejected that out of hand on the grounds of my own personal safety. Things were pretty rough there then and I decided I wasn't even going to ask for a visa."

Tassan made arrangements to fly to Johannesburg in April, spend five weeks collecting in South Africa, and then go on to southern France for another week. The entire trip was to cost him 4,000 dollars, most of it for transportation. In Johannesburg he rented a Ford Escort and drove to the capitol at Pretoria.

"I was extremely disoriented," Tassan remembered. "It had been a very turbulent flight across the Atlantic, and at first we couldn't land at Johannesburg because of thick fog. I had a very poor sense of direction, which is unusual for me, but it lasted most of the time I was in South Africa. I attributed it to being below the equator for the first time and having the sun in the wrong place. Of course, having the steering wheel on the wrong side of the car didn't help, either."

But the drive to Pretoria through the dry landscape of the antipodal autumn did much to reassure him. "It was exactly like the drive from Berkeley to Sonoma or Santa Rosa, with rolling hills, dry grass, pines, and all that. I had to pinch myself to realize I wasn't back in California. The only thing really different was the large number of pedestrians along the highway, especially the women carrying loads of wood."

In Pretoria, Tassan went straight to South Africa's equivalent of the U.S. Department of Agriculture, the Plant Protection Research Institute. There he was given much helpful information by two entomologists, D. P. Anneke (now deceased) and Gerhard Prinsloo. They provided him with maps and traveling tips and introduced him to the resident expert on ice plant scales, G. De Lotto. Tassan showed De Lotto the specimens of scales he had brought with him and mentioned the dawning realization in California that they were dealing with two species.

" 'Oh, yes, we have them both here,' he told me, but he hadn't gotten around to publishing a description of the new species yet," Tassan said. "Then he took me outside and showed me a small bed of ice plants in the courtyard where one of the natural enemies was working over the mature female scales like crazy."

Within a few days, Tassan was on his way, following his new friends' instructions on where to look for ice plants. His route took him southeast through subtropical Natal to the warm coast of the Indian Ocean, north of Durban. From there he drove the length of the coast, with occasional forays inland, through Durban, Umtata, East London, Grahamstown, Port Elizabeth, Plettenberg Bay, Swellendam, and Stellenbosch to Capetown; from the southwestern coast he returned by way of Bloemfontein to Johannesburg. Much of his search, of course, was on foot, poking about in waste places and walking miles of beaches.

"The local people have a variety of vernacular names for the plant, like Hottentot fig," Tassan said. "The plant was not nearly as common as I thought it would be and sometimes I'd walk two or three miles along a beach and find just one little patch, covering a couple of square feet. And once I found ice plants, the scale was really hard to locate. The white egg sac was quite obvious, but in its immature stages the scale is greenish and easy to overlook. So one technique I used was to look for movement—the movement of ants on the leaves, because they tend the scale for honeydew."

The search demanded intense concentration in order to pick out the plants and then the scales from the teeming array of living things in an unfamiliar landscape. It was especially difficult to focus on the job at hand because the brilliantly colored birds and butterflies constantly seemed to cry out for attention.

"I don't think I collected more than 50 specimens of other insect species on the whole trip," Tassan recalled. "I was focusing too hard on ice plants. But I'll never forget one insect I found. I did make it a point to collect ants on ice plants so that I could write a report on their association with the scale. One day near Port Elizabeth I spotted what I thought was an ant. It scurried around on the leaves and even seemed to stroke the scale to get it to produce honeydew. So I popped a few of them into alcohol to bring back to California with me. I was in for quite a shock."

But that was delayed until his return to the U.S. On another occasion, south of Durban, he found a beetle enthusiastically attacking the scale. It seemed so effective, and reminded him so much of the behavior of the vedalia beetle, that he cabled his superior, Kenneth S. Hagen, in Berkeley: "I've found it." But later, when it was released in California, Tassan's South African superstar failed to establish itself.

"One of the things that kept me quite alert in the field was the possibility of poisonous snakes," he said. "I had asked Gerhard Prinsloo in Pretoria about snakes, and he discounted my chances of having any trouble. But he did mention a couple of mambas and cobras whose bite was so deadly that the chances of recovery were almost nil. When I told him I had brought a snakebite kit with me, he found it very humorous. 'That's not going to do you any good if one of *those* snakes bites you,' he said. So, there were areas where I saw ice plants but I didn't investigate because of the dense surrounding vegetation. It turned out that the only snakes I saw in South Africa were at a snake farm which I visited to get an idea of what they were like. They were *very* impressive."

Tassan collected nearly 10,000 individual specimens of predators and parasites (many of them in their parasitized hosts). His collection represented a wide variety of locations, both in the wild and in cultivated stands, from subtropical Natal to the temperate region around Capetown. Two of the most successful parasites proved to be small wasps, *Metaphycus funicularis* and *Metaphycus stramineus*, which deposit one or more eggs in an immature scale. Both these parasites reproduce quickly, as their life cycles are limited to two or three weeks. Tassan also collected a successful predator, *Exochomus flavipes*. This species of ladybird beetle had been introduced to California earlier to combat another scale, but was unable to establish itself. The ice plant scale seems to have been the tonic it needed to make good in northern California.

Tassan's trip was often arduous. During the day he collected specimens in vials and other containers and kept them in a styrofoam ice chest in the back of the car. When he reached a hotel at the end of the day, he would ask the manager to freeze the dry ice so that he could use it in the chest again the next morning. Then he would clean up whatever notes or cataloguing had to be done for the day.

On weekends, Tassan packed the specimens for shipment to California. The idea was to ship the first thing on Monday mornings so that, if there were any lengthy delays, the shipment would not arrive at the end of the week and remain at the airport all weekend because there was no one on duty at the university to pick it up. In fact, only one shipment was ever delayed. That one was somehow stranded in Chicago for more than a week, but illustrating the tenacity of insect life, the parasites survived. Ordinarily, a shipment reached Berkeley within 36 hours.

Tassan had brought vials and other small packing materials with him from California, and the entomologists in Pretoria provided him with shipping boxes. He was especially careful in packing the parasitized larvae, putting honey into their boxes so that the parasites would have food if they emerged on the flight to California.

"I had arranged for the various permits needed to ship insects out of South Africa and into the United States," Tassan said. "The most reliable way to send my specimens was through a shipping agent, who would make certain that the material got aboard a plane quickly. Then I would call Berkeley myself and arrange for someone to be at the airport and shepherd the shipment through USDA and customs. The packages, of course, were not supposed to be opened at the airport, so they could be brought right to our own quarantine at Berkeley."

At the end of five weeks, when Tassan's wife and eight-year-old son joined him in Johannesburg, the family spent a week sightseeing in Kruger National Park. They went on to southern France, where Tassan collected for another week.

"It was the wrong season," he admitted. "The scale had just reproduced and they were too young to be parasitized. I hand-carried some scale back to California, but there was nothing there."

There was something, however, in Berkeley when he returned. He had shipped home the seed for sowing a rich crop of natural enemies against the ice plant scale throughout California. They were bred in the university's insectary and the offspring were distributed by the thousand through the state's ice plant beds. Caltrans, having given the Division of Biological Control less than 300,000 dollars in grants over five years, looked forward to the permanent control of the scale in return for its investment. In the last couple of years, spraying insecticides for the ice plant scale has been almost completely confined to counties where parasites have not yet become established. A dividend since Tassan's return has been the addition of another parasite to the arsenal.

"It took me a long time to get around to looking over the ants I had collected," Tassan said. "A year or so after I got back we received a shipment of a brand-new parasite, a little wasp called *Encyrtus saliens,* that the entomologists in Pretoria had discovered. I thought it looked familar. I checked my notes and then took out the vial in which I had kept what I thought were the ants I found near Port Elizabeth. Sure enough, my 'ant' was the little wasp *Encyrtus.*"

Encyrtus, he learned, is a solitary parasite whose wingless females somewhat resemble ants. It is now living in the wild in northern California.

"Of course, when I took a good look, I could see quite plainly that it was a wasp," Rich Tassan said, grinning. "For my own self-esteem, I just hope it doesn't turn out to be the most *important* parasite on the ice plant scale."

9

THE DEVASTATING SCREWWORM

The Florida Strategy

THE SCREWWORM (*CALLITROGA HOMINIVORAX*), WHICH ONCE ranged over vast areas of the southern United States, is looked upon by humans as not only one of the most destructive but also one of the most repulsive creatures in the world.

Its specific epithet, *hominivorax,* means "man-eating," and suggests one of the reasons why it is abhorred wherever it appears. The screwworm is one of the blowflies, a distant relative of the familiar bluebottle and greenbottle flies of Europe. Like those of other blowflies, its larvae eat flesh. But whereas many members of its family feed mainly on carrion (bluebottles and greenbottles are traditional pests of the kitchen, where they lay their eggs in stored meat), the screwworm attacks living flesh. The wounds of soldiers during the American Civil War are said to have been occasionally infested by this insect. Yet its most virulent assaults are against wild and domestic animals. Ranchers can think of no more distasteful sight than that of hundreds of screwworms lodged tenaciously in a wound or a sore around the eyes or mouth of one of their prize cattle.

The adult screwworm, a two-winged fly of metallic blue, is about three times larger than a housefly. The female seeks out any area of raw living flesh where blood is at the surface, including scratches and insect bites. Domestic sheep nicked during shearing and the navels of newly born animals are especially vulnerable. The female lays a compact mass of 200 or 300 eggs in the wound. (*Blowfly* comes from the Old English word "blowing" as used to describe a fly depositing its eggs, and thus maggoty meat was said to be "fly-blown.") Edward F. Knipling, who studied the screwworm throughout most of his career in the USDA, has graphically described the aftermath.

"Tiny maggots hatch from the eggs in 12–24 hours," Knipling wrote. "They begin feeding on the flesh head-down and closely packed in the wound. The feeding larvae cause a straw-colored and often bloody discharge that attracts more flies resulting in multiple infestation by hundreds to thousands of maggots of all sizes. Death is inevitable unless the animal is found and treated."

After about five days, the larvae have reached a length of two-thirds of an inch. They drop from the wound, burrow into the soil, and quickly pupate. There the larvae body dissolves, eventually rebuilding during the miracle of metamorphosis into the adult fly. The screwworm fly, which lives for two or three weeks in summer, soon mates and spends the rest of its life feeding on nectar, pollen, and carrion. (Like the housefly, it uses its proboscis to inject solid foods with chemicals that liquefy them.)

The screwworm may go through a generation in as short a period as three weeks. In warm regions where the insect survives the year round, there may be from 10 to 12 generations annually.

"We stand now where two roads diverge," Rachel Carson wrote in *Silent Spring* as she surveyed the impending choice between using more insecticides and finding alternatives to solve our pest problems. "But, unlike the roads in Robert Frost's familiar poem, they are not equally fair. The road we have long been traveling is deceptively easy, a smooth superhighway on which we progress with great speed, but at its end lies disaster. The other fork of the road—the one 'less traveled by'—offers our last, our only chance to reach a destination that assures the preservation of our earth."

The first of these "other roads" she chose to write about in her book was the sterile male technique developed by Edward Knipling, who became chief of the USDA's Entomology Research Branch in the 1950s. It was Knipling's ingenious idea to rear pest insects in enormous numbers, sterilize them through radiation, and then release them to compete with normal males for females in the wild. In this "autocidal" campaign, the insect spreads sterility, and eventually eradication, through its own population. The control Knipling was able to achieve over the screwworm with this technique was compared at a recent scientific symposium "with the completion of the railroads into the West and the fencing of the open range in its impact on beef production methods in this country." The technique fulfilled Rachel

Carson's high hopes in almost completely eliminating the need for insecticides against a major pest, and it even raised expectations in other quarters that it could be successful against other pests.

"Theoretically it can be shown that the relative effect would be equally or more dramatic when applied to higher animals," Knipling wrote in *Scientific American.* "Under certain circumstances it may be desirable to eliminate noxious animals such as rodents, predators, destructive birds or aquatic animals from certain areas, or merely to regulate the population of desirable animal species, such as certain large predators or large game species." Elsewhere in his article he added: "No species can acquire immunity to sterile matings as it can to the insecticides that have been used in the past."

Knipling's expectations, as we shall see in a later chapter, were too high. It has become one of the self-evident truths of pest control that the complexity of the ecosystem rules out a single panacea for all problems. Knipling was even wrong in believing that the screwworm could not acquire immunity to sterile matings. Yet his sterile male technique is undeniably one of the outstanding triumphs of applied science ever achieved in biology, for no other eradication project has succeeded over such a vast area.

The screwworm apparently originated in the American tropics and may have invaded what is now the United States no more than 300 or 400 years ago. An important aspect of the screwworm's history, and one that ultimately proved decisive in its control, is that it never existed in the staggering numbers characteristic of such wide-ranging southern pests as the fire ant and the cotton boll weevil. In the United States, it was mostly confined to southern Texas, representing a spillover population from the center of its range in Mexico and Central America. It has, however, a considerable capacity for flight; adults released into the wild have been recaptured as far away as 180 miles. As the weather warmed in summer, the adults tended to fan out from southern Texas northward into the Midwest, and a few might be found as far north as the Canadian border. But cold weather always killed off this advance guard and only the screwworms in southern Texas managed to survive the winter.

The building of the great livestock empires in Texas during the latter part of the nineteenth century brought the screwworm out of obscurity. The land had been transformed. Windmills appeared everywhere on the Texas horizon, bringing water to the surface and creating waterholes for cattle. The new railroads linked the state to markets in the Midwest and East. The screwworm, subsisting earlier on the occasional wild animal, or scattered cattle on pioneer ranches, suddenly found itself with an unlimited food supply on range land densely covered by sheep and cattle. University of Texas entomologists R. H. Richardson, J. R. Ellison, and W. W. Averhoff have summarized the events that created a serious pest.

"The great increase in cattle led to increasing outbreaks of screwworms which, by the end of the century, were becoming an annual curse on ranches in Central Texas and along the Gulf Coast," they wrote in *Science*. "It also led to severe overgrazing and fewer prairie fires and, consequently, changes in the distribution of certain plants. For example, the chaparral in southwestern Texas and Northeastern Mexico became a thorn forest of woody legumes, as did the central and southern Gulf Coast regions of Texas. Deer replaced the antelope as the dominant game animal as cattle had replaced sheep. Since screwworm flies feed on the nectar of such legumes and since, like cattle and deer, the flies are associated with water courses, the flies, cattle, and deer

were generally in close proximity and screwworm populations could rapidly expand."

Much larger populations of screwworms became established in other parts of the South, spreading quickly in
warmer weather either by their own powers of flight or on
infested cattle shipped from one region to another. By 1933,
screwworms had reached Florida and Georgia. They were a
terrible plague wherever they became established. Livestock
owners reported about 1 million cases of infestation among
their animals each year, certainly but a fraction of the actual
toll.

"A particularly disgusting and sickening job was when
cows or calves got screwworms in their mouth and gums," a
producer told the historian C. G. Scruggs. "This came about
in two ways. One, the cow or calf—if they could reach the
wound—would try to lick the worms out of the lesion. Thus,
some live worms would get in the mouth of the animal and
take hold. In some cases, I'm sure that flies would also lay
eggs in the mouth of the newborn calves. A small calf—
when sleeping—will let its mouth hang open. Thus the
moist gum and lip areas were a target. You couldn't use any
medicine—just remove the worms and hope you got them
all. Some cases would be so bad that an animal might lose
some of their teeth. It sure wasn't a job for anyone with a
queasy stomach."

The screwworm was a costly burden for southern livestock owners. Ranching became especially labor-intensive
because an untreated animal was a doomed animal. Entire
herds had to be checked every few days for infestations.
Most remedies proved worthless, although eventually a con-
coction of benzol and pine tar oil was found to be successful
in treating infested wounds. Yet the screwworm disrupted
an entire industry. Calving had to be restricted to the winter

months, for in warm weather screwworms attacked the navels of the calves as well as the genitals of their mothers unless repellents were smeared on the vulnerable areas. Branding, de-horning, and castration also had to be crowded into a few winter months. Interrupted production thus brought all sorts of economic hardships to both ranchers and cowboys (the latter being laid off for months at a time). The effects on the larger wildlife, especially deer, were catastrophic because infested fawns inevitably perished.

It was in this climate of economic hardship that Edward Knipling grew up on a farm along the Texas coast in the first quarter of this century. He watched his father, a German immigrant, struggle hopelessly against the boll weevils that destroyed his cotton crop and the screwworm that attacked his livestock, and the boy himself helped doctor the infested animals. Understandably, the young Knipling wanted no part of a farmer's life.

"Knipling was not a collector of insects as a boy, but he liked to watch their behavior during his youth in rural Texas," wrote John H. Perkins in his book, *Insects, Experts, and the Insecticide Crisis.* "His initial schooling was difficult, partly because he had to become Americanized from his German-speaking homelife, and partly because competition was keen for rural children when they entered the town school in Port Lavaca. He persevered, however, and ultimately completed his bachelor's degree in entomology at Texas A&M University in 1930. He was the only boy in his family to pursue a higher education, and he continued it through his master's (1932) and doctoral degrees (1947) at Iowa State University. As with most students, his future profession was in no way foreordained. He knew he didn't want to pursue farming as a career; hence, he went to college. He enjoyed genetics as an undergraduate, but he chose entomology because of more potential employment. Pragmatic considerations in Knipling's case produced a happy match between personal interest and opportunities for work."

After receiving his master's degree, Knipling went to work for the USDA, where he remained for the rest of his career. He advanced rapidly in the agency, working mainly on the screwworm, as well as other insects harmful to humans or domestic animals. During World War II, he was assigned to a laboratory in Orlando, Florida, where much of the early work on DDT (then being developed for wartime use against lice and malaria-carrying mosquitoes) was carried on. He shared the enthusiasm for the new synthetic chemical insecticides. These substances seemed so effective that entomologists began to contemplate the total eradication of major insect pests.

It was with some dismay, then, that Knipling became aware of the growing resistance to the new chemicals among mosquitoes, house flies, and the boll weevil. The dream of permanent control was fast fading. But Knipling, intelligent and imaginative, had been pondering alternative methods for some time, although his investigations had been sidetracked by the war and the elation over the early success of the new chemicals. Extremely thoroughgoing and perhaps fired by his family's experiences with the despised screwworms and boll weevils, he was impatient with the imperfect control often exerted on pests by the use of natural enemies. He yearned for nothing less than the complete extermination of the insects he considered serious threats to the welfare of mankind. Another approach had been taking shape in his mind for a long time.

"In 1937 it occurred to me that it might be economically feasible to rear and release screwworm flies in sufficient numbers to exceed the natural population," he recalled afterward in *Scientific American*. "The idea was suggested, in fact, by the observation of my colleague A. W. Lindquist, of the Uvalde, Texas, Station of the Department of Agriculture, that the number of screwworm flies trapped during the winter in that region is exceedingly small. If some method could be devised to cause the artificially reared and released insects to destroy those in the natural population, I thought, this might provide a means for annihilating the insect. The development of a genetic strain carrying a factor that would be lethal under natural conditions was one possibility. Development of a chemical that would induce sterility in the flies before release was another. However, the most promising approach seemed to be the sterilization of the flies by X- or gamma-rays. As long ago as 1916, G. A. Runner of the U.S. Department of Agriculture had shown that cigarette beetles produce infertile eggs after exposure to X-rays, and at the University of Texas in 1928 H. J. Muller had demonstrated similar effects in fruit flies. Significantly for my purpose, it had been found that such exposure at the right stage in the insects' development had no other adverse effect."

Like many other extremely imaginative ideas, Knipling's were not received enthusiastically within the government. But by 1950, he was one of the USDA's most influential and respected research scientists. The biology of the screwworm had been worked out. (Until 1933, it had been confused with a carrion-feeding blowfly so that entomologists believed the best way to fight it was to quickly dispose of all carcasses on rangeland.) Other scientists had figured out a way to rear screwworms for experimental purposes in the laboratory on a diet of raw meat. Thus in 1950, Knipling asked the advice of H. J. Muller, who had been awarded the 1946 Nobel Prize in Physiology and Medicine for his work in artificially inducing gene mutations in fruit flies (*Drosophila*) by means of X-rays. Muller was supportive, expressing confidence in Knipling's plan, provided that sterilized male screwworms were able to compete successfully for females in the wild.

The USDA began research at a laboratory in Kerrville, Texas. There it was determined that sterility could be induced in male screwworms by exposing the pupae to 2,500 roentgens of ionizing radiation. Although this, or even higher levels, caused dominant lethal mutations in the sperm, the insects apparently suffered no other disability. Females became sterile at over 5,000 roentgens.

"Sexually sterile males were then placed in cages with normal males to test their ability to compete with normal males in mating with normal females," Knipling wrote. "The results were extremely favorable. When the ratio of sterile to fertile males was one to one, about half the normal females produced sterile eggs. When the ratio was stepped up to nine to one, the sterility in the females was 83 percent, sufficiently close to the theoretical expectancy of 90 percent. The investigators showed that females of the screwworm fly mate only once and do not discriminate between irradiated sterile males and normal males. The presence of irradiated sterile females in the caged population did not alter the results, which meant that it would not be necessary to separate the sexes before releasing the flies in the field."

Knipling's mathematical models had predicted that, by releasing nine sterile males for every one male in the field, entomologists would be able to eliminate the wild fly population in five generations. The laboratory tests suggested that those projections were accurate, but he could never be certain until the program was tested in the field. He knew of a promising site for an experimental release of sterile males. Sanibel Island, an area of about 15 square miles off the west coast of Florida, had a flourishing screwworm population. Under the supervision of Knipling's colleague, A. H. Baumhover, and working with flies irradiated by Raymond Bushland at the USDA laboratory in Texas, the researchers carried out a three-month experiment on Sanibel. At the end of that time, the screwworm had virtually disappeared from the island, although Knipling was not able to demonstrate eradication because a few wide-ranging mated females continued to fly to Sanibel from the mainland, only two miles away.

In the early 1950s, the old Bureau of Entomology was abolished and replaced by the Entomological Research Service, with Knipling as its chief. He was now the highest-ranking research entomologist in the USDA, a position from which he exerted considerable leverage to pry more money from the budget for his sterile male program. In 1953, he was presented with an opportunity to prove that entomologists now had the tool to eradicate an entire population of screwworms. A veterinary officer on Curacao in the Netherlands Antilles approached Knipling and asked his advice about controlling screwworms that were attacking the island's goats. When Knipling located the island on a map and saw that it was large (170 square miles) and separated from any other land by at least 40 miles of sea, he became keenly interested in the veterinarian's problem. After brief negotiations with Dutch officials, he obtained permission for a team of USDA entomologists to try to eradicate the pest on the island.

This time, Knipling and his colleagues arranged to rear very large numbers of screwworm flies at a laboratory in Orlando, Florida. To irradiate them more efficiently, they borrowed a cobalt-60 gamma-ray source from the Oak Ridge National Laboratory, then had the flies shipped by commercial airline to Curacao. Crop-duster planes released them over the island at the rate of 400 sterile males per square mile each week. The success was more dramatic than anyone had earlier hoped. By the fourth generation, the screwworm had been extirpated from Curacao.

Knipling was now ready to tackle the Florida mainland. The state matched federal funds to convert a hangar into a screwworm-rearing plant at an airport in Sebring. The program was in full operation by the summer of 1958.

"The insects received the best of care in an air-conditioned, humidity-controlled room," Knipling wrote of the new plant. "Thousands of flies roosted on cloth strips in each of a large number of screen cages. They were fed a diet of honey and extracts of ground meat, and after about eight days were induced to lay their eggs in a slurry of ground meat, blood and water in which pupae had previously been raised. The larvae generated factors in this medium that proved highly attractive to females ready to deposit their eggs, especially when the medium was heated to the body temperature of livestock [approximately 100° F].

"After hatching and five days of feeding, the larvae crawled out of the feeding vats into a large funnel, where they were collected in trays of moist sand. There they changed to pupae in about 24 hours. Screened from the sand into special screen baskets, the pupae were held at a controlled temperature of about 80 degrees. On the sixth day they were placed in cylindrical metal containers and exposed to 8,000 roentgens of gamma radiation from a cobalt-60 source, enough to assure the complete sterility of both male and female flies.

The production line required six 500 curie cobalt-60 sources. After irradiation the pupae were packaged in small cardboard boxes in readiness for delivery to the field when they emerged as adults two days later."

This was an insect control program of a scale and complexity never before imagined. During the next year, the screwworm larvae reared in the laboratory consumed more than 40 tons of ground meat, most of it the flesh of horses and whales. Each week, air-conditioned trucks delivered 50 million screwworms to a fleet of aircraft, which at peak periods numbered 28 single-engine planes carrying mechanical equipment to break open the cardboard boxes of flies and release them on designated routes. In all, more than 2 billion sterile flies were dropped within 18 months over 70,000 square miles of Florida, Georgia, and Alabama. Incredibly, Knipling's plan was a complete success. The Sebring plant was shut down in the fall of 1959, the screwworm having been wiped out in the southeastern United States.

But ahead lay the vast proving ground of Texas, separated from that bottomless reservoir of tropical fauna in Mexico by one of the most permeable of the world's political borders. Once more the belief that man can single out an opportunistic alien species for complete domination was to be badly shaken.

10

THE DEVASTATING
SCREWWORM
The Texas Solution

ALMOST NOTHING WAS KNOWN ABOUT SCREWWORMS IN THE wild, until recently. In this respect, the program to control the screwworm by the sterile male technique is reminiscent of the equally phenomenal success entomologists enjoyed in controlling the cottony-cushion scale with the vedalia beetle. Scientists, in both cases, were fortunate in finding a perfect match between the pest and the means of controlling it. But a major difference also exists in the two programs. The scale has not yet found the means to escape from the dominance of a natural enemy. The screwworm, on the other hand, is dealing with an "unnatural enemy"—members of its own species fundamentally altered by a sophisticated instrument devised by man. Entomologists who directed the screwworm campaign were on thin ice when they failed to investigate every facet of their target insect's behavior in the wild.

In the absence of concrete information about the responses of a species to new conditions, scientists often employ a surrogate species. Thus, researchers have been able to extrapolate the results of experiments on mice or rats to the higher mammals, including man. The information accumulated by such extrapolations is not exact, but it may provide valuable clues to the solution of a problem. Two researchers at the Unversity of Texas, Merry Eve Makela and R. H. Richardson, have performed studies on fruit flies that may give pest control specialists important clues in keeping the lid on screwworm populations.

Fruit flies, especially *Drosophila melanogaster*, are ideal laboratory subjects for the study of genetics. They breed in large numbers, complete their life cycles (from eggs, larvae, and pupae, to the laying of the next generation of eggs) in less than two weeks, and feed readily on rotten fruit. Scientists are able to maintain them with little fuss in milk bottles or other discarded containers and, because fruit flies are attracted to light, lure entire colonies from one container to another with a simple light source. Finally, fruit flies possess large, easily studied, chromosomes, most notably in the salivary glands of the larvae, a characteristic that endears them to geneticists.

Makela and Richardson studied the genes (arranged linearly on chromosomes) in what appeared to be a single interbreeding population of fruit flies from a specific geographical region. With the aid of gel electrophoresis, a sophisticated technique that separates proteins on the basis of their total electric charge, they were able to locate and analyze simultaneously as many as 30 or 40 of an individual fly's genes. In this way, the two scientists discovered genetic variations, hitherto unsuspected, in the population. They found they could separate the laboratory population, which on physical characteristics seemed to be a unit, into subpopulations distinguished by very different arrays of genes.

Once they had the subunits identified, Makela and Richardson learned (through the scarcity of genetic hybrids appearing in their electrophoretic studies) that there was little interbreeding among the subunits. These populations were isolated from each other not by geographical boundaries, but rather by their mating habits. They were representatives of types that were already evolving into new species in the wild. Other scientists were uncovering similar "cryptic," or hidden, species among what had been thought to be interbreeding populations.

Because the sterile male campaign had moved closer to the heart of the screwworm's range in the tropics, Makela and Richardson knew that the studies on fruit flies might yield results that could be extrapolated to the screwworm: A variety of experiments had disclosed that cryptic species tended to increase in the center of an organism's range. If control specialists did not recognize genetic differences in their factory-reared population and make certain that all strains were preserved for release into the wild, there could be trouble ahead. Females in certain strains of the wild population might not accept the "alien" strain of sterile males.

"When the laboratory culture containing only part of the strains or species is released in a pest control effort," Makela and Richardson wrote, "only some of the pests are decreased while their uncontrolled relatives are free to invade a now-empty niche. If two species were competitors, they may have served to restrict each of the other's ranges. Upon removal of the competition, a remaining species may extend its range to include both new geographic areas and new foods formerly confined to the now-decimated species. *Unless both species are recognized and suppressed or eradicated simultaneously, the control program may fail.*"

By the early 1960s, Texas stockmen, well aware of the USDA's extraordinary achievement in the Southeast, were eager to begin their own final offensive against the screwworm. Their labor problems were becoming increasingly acute as ranch hands, discouraged by their unemployment when most livestock operations during the screwworm season were curtailed, had deserted the range for permanent jobs elsewhere. More than 100,000 southwestern producers contributed money (based on how many heads of livestock they owned) to a nonprofit foundation to help pay the program's cost.

In 1962, the USDA built a new mass-rearing facility at Mission, Texas, designed to produce 150 million sterile flies a week. Perhaps the chief bottleneck in rearing any insect on a grand scale is food. Raw meat and whole blood, which had been the mainstays of the screwworm program for some years, were costly and now presented the laboratory staff with almost insurmountable problems in procuring and storing the needed quantities. Once more the staff was up to the challenge. Its scientists developed an artificial diet for screwworm larvae consisting of spray-dried blood, powdered cottage cheese, powdered egg, cotton, and calf suckle. The early results were disappointing because microorganisms quickly fermented the mixture, and the larvae failed to grow as large as those on the original diet. But when the staff added small amounts of formaldehyde to the mixture, fermentation did not occur and the larvae reached normal size.

The USDA began the program by covering 170,000 square miles of Texas and New Mexico rangeland with sterile males. Mexican authorities cooperated with the agency, permitting the planes to release flies in a band that extended 50 miles south of the border, thereby establishing a buffer zone. To accommodate the larger number of flies, the staff switched to bulkier cartons, each with a capacity of 2,000 flies instead of the earlier 400 flies. The USDA also discarded the smaller aircraft used in Florida and bought twin-engine Beechcraft and DC-3s to provide greater range and capacity.

The USDA had embarked on what was, in effect, saturation bombing. The screwworm population, especially in hot, dry weather, may exist at very low densities early in the year, perhaps from a single insect to 200 an acre. Thus, the program's success depended on both the quantity and quality of its sterile flies. Enormous numbers of strong-flying males would be necessary to find the few females existing in the wild, for only when the sterile males pass seminal fluids to them will the females stop mating and begin to lay their eggs. Infertile eggs, even when laid in animal wounds, inflict no further harm on the victim. The program's string of spectacular triumphs continued in the Southwest. What had once been 1 million or more annual screwworm infestations reported among southwestern livestock declined to 51,592 in 1962 and 7,111 in 1963. In 1964, the total plummeted to 395—almost half of them in Arizona, where releases had just begun.

The releases were coupled with an educational campaign to persuade ranchers to treat their animals' infested wounds with benzol and pine tar oil and thus destroy whatever fertile eggs had slipped through the barrage of sterile males. Extension specialists advised producers to be on the lookout for a rise in screwworm populations in their area so that they could delay ranch operations, such as branding and de-horning, that might cause a large-scale infestation. Complete eradication obviously had not been achieved, as small numbers of the pests continued to invade the United States from Mexico. But reported infestations amounted to only a few hundred a year during the rest of the 1960s and early 1970s, except for a flare-up in 1968 that entomologists attributed to the ranchers' carelessness. Producers, faced with a labor shortage in an inflationary era, had been saved hundreds of millions of dollars. (According to rather sketchy figures compiled from various sources in the 1980s, American stockmen have saved 2 billion dollars in stock losses and labor costs, at least 113 times the total funds spent on the sterile male program.) A twist to this story is that the Texas deer population, suppressed for so long by its losses to the screwworm, has exploded in recent years to cause some agricultural damage of its own.

Then, in 1972, something went wrong. Mild weather permitted a number of screwworms to overwinter successfully in southern Texas and northern Mexico. A flourishing deer herd and relaxed vigilance by producers gave the insects a head start in the spring, despite the fact that the mass-rearing facility began to turn out more than 200 million flies a week.

"Screwworms were multiplying in record numbers in South Texas and were spreading to all areas of Texas by mid-April," entomologist James E. Nory wrote afterward. "By mid-May the screwworm situation was completely out of hand. Many feel that if maximum production had been achieved earlier it would have had little effect except to possibly help slow the spread. The population of screwworm flies migrating north from Northeast Mexico was too large to be sufficiently outnumbered by the production and distribution of sterile flies."

There were nearly 100,000 infestations reported by producers that year, most of them in Texas. Explanations that emphasized the mild winter or the producers' carelessness did not seem adequate in view of this tremendous surge. Although the USDA kept the situation from getting out of hand, the agency during the next few years was not able to bring the screwworm population down to pre-1972 levels. It was apparent that the sterile males produced at Mission were not performing as efficiently as they had in the program's early years, even though they were being released in larger numbers. Why had the program bogged down?

Scientists working with insects as various as fruit flies and codling moths had known for some time that species mass-reared in the laboratory may undergo rapid genetic changes, either by adapting to unusual conditions in confinement or by a series of random mutations called "genetic drift." In an attempt to forestall such occurrences, the staff at the Mission facility brought fresh strains of screwworms, collected at different places in Texas, Mexico, or the Caribbean Islands, into the laboratory and phased out older strains. Nevertheless, the factory-reared flies proved to be smaller than their wild counterparts, while showing other minor differences in behavior, eye color, and flying ability.

In response to the new crisis in screwworm control, scientists put forth several theories. They suggested that the wild and factory-reared flies were evolving simultaneously in different directions. Conditions in their confinement may have triggered adaptations among the Mission strains that unfitted them for life in the wild. At the same time, the wild variety may have adapted to the increasing pressure represented by the 10 billion sterile flies poured into the environment every year; in that case, the wild flies may have learned to discriminate against their irradiated look-a-likes.

The USDA began a cooperative research project with the University of Texas at Austin to determine how the screwworm threatened to foil this sophisticated pest control program. The results demonstrated once more the endless capacity of insects to adapt quickly to changing conditions. Guy L. Bush and two colleagues at the University of Texas described conditions at the Mission facility in 1972, where screwworm adults were held in large cages in total darkness at 77° F.

"The mass rearing methods in use were designed to discourage flight activity in order to reduce damage to the parental flies and increase egg production," they wrote in *Science*. "Each cage was sub-divided by many large sheets of newsprint suspended from the cage ceiling to increase the surface area and to inhibit movement. Food and oviposition sites were placed on the floor of the cage. A premium was therefore placed on the fly that walked rather than flew or that flew only short distances. In addition, these flies were reared at a relatively high constant temperature and were thus not exposed to the highly variable weather conditions encountered by released individuals in nature. Selection for maintaining adequate flight muscle activity to ensure normal dispersal in nature was therefore relaxed. A further premium was placed on rapid development from egg to pupa at high constant temperatures; short-generation flies were positively selected in order to promote efficient plant operation regardless of their overall suitability as competitors against wild flies."

The researchers in Austin decided to concentrate on genetic changes that may have taken place in the factory flies, especially in flight muscles that are known to be affected in artificially reared populations of other species. Using gel electrophoresis, they were able to separate proteins in the enzyme that controlled the flow of energy through the flight muscles. The arrangement of amino acids in these proteins determines the biological properties of each molecule. "These differences reflect genetic variations in the encoded messages stored in the gene," Guy Bush wrote. "Therefore, if one can identify molecular variants in a specific protein, one may estimate the genetic variation existing in the gene that specifies the protein."

Bush and his fellow researchers discovered that the factory flies differed from wild ones in genes that controlled their flight muscles. Further studies showed that the enzyme regulating energy flow was comparatively inactive in the lower ranges of the temperatures the flies were likely to encounter in the natural environment. These strains, in effect, had adapted to the high temperature maintained in their cages.

"Although the mating behavior of wild flies has never been observed in nature," Bush wrote, "there are some reasons to suspect that at least part of the courtship of this fly occurs in the air or at specific sites which require normal flight activity. Thus, the factory-reared males would be at a considerable disadvantage in competing for mates."

Bush's findings were soon confirmed in the wild. A team of USDA researchers examined all screwworm flies that came to lay their eggs in wounded animals kept for experimental purposes. The researchers were able to distinguish the wild flies from the factory strains because of their larger ovaries. They found that the wild flies began arriving on the experimental animals early in the morning and remained active all day. The factory flies, however, did not seem to be able to fly until the mid-day temperature had warmed their flight muscles. Although male flies were excluded from the study because they do not visit wounded animals, the researchers were satisfied that the sterile males shared the sterile females' inability to get moving early in the day. The normal males were likely to find the normal females first, thus ensuring a substantial supply of fertile eggs.

These obstacles to a successful control program could be overcome by refining the methods of rearing flies in the factory. The staff members at Mission, indeed, began to monitor genetic changes in their fly populations more closely. For instance, before introducing a new strain to the facility, they exterminated the previous breeding stock to make certain it did not contaminate the newcomers with undesirable genes. But are there other, more subtle, genetic changes occurring in factory strains that have not yet been detected? The lack of information about many details of screwworm behavior in the wild makes exact comparisons of wild and factory strains almost impossible. Meanwhile, are the wild insects adapting to the pressures that entomologists have put on them? In other words, has the screwworm developed resistance to the sterile male program as other insects developed resistance to DDT?

After another screwworm outbreak in the Southwest, Richardson, Ellison, and Averhoff, the three University of Texas entomologists mentioned earlier, intensified their genetic studies of screwworms captured in the field. They discovered a number of different genetic types of screwworms in both the United States and Mexico. (Although the scientists believed that some of these populations may eventually be considered different species, at that point they preferred to call them "types.") The ranges of these types often overlapped, but they did not interbreed and were, thus, reproductively isolated. Variations in chromosomes among the types affected such morphological features as the structure of the male genitalia, and such behavior as the patterns of oviposition. Wounds infested by screwworms generally found in the United States give off a characteristic odor so that cowboys are able to locate infested animals that may be hidden more than 100 yards away. The entomologists could detect no such odors at close range from wounds infested by several types of screwworms found in Mexico.

Why, then, was the sterile male program so successful at first? Richardson and his associates believed that during the early years the program's breeding stock originated in the same populations into which the factory-reared flies were later released. But when the native wild populations were suppressed, other flies invaded the Southwest from Mexico, and these invaders differed genetically. They were not as likely to accept as mates the sterile males from the original types. Thus, proportionately fewer clusters of infertile eggs led to screwworm outbreaks.

Recent developments have emphasized the potential threat of this sort of insect "resistance." The grand design for the program as envisioned by Edward Knipling (who retired from the USDA in 1970, although he remained active as a consultant and writer), is to gradually push the screwworm population south through Mexico and eventually to Panama. Through an agreement between the United States and Mexico, a new factory was built near Tuxtla Gutierrez in the Mexican state of Chiapas, and the Mission rearing facility was closed in 1981. The Mexican facility produces 500 million sterile flies a week, some of which are available to carry on the program in the United States. However, as the area covered by the program increases and new types of screwworms are encountered, the "match-ups" between sterile and normal flies may become increasingly uncommon. Entomologists are also remaining alert to the development of parthenogenesis (which has been noted elsewhere) in isolated screwworm populations. Some female insects, in the absence of fertile males, have the capacity to lay fertile eggs—in other words, a capacity for virgin birth. Such a development, of course, would put that population beyond control by the sterile male technique.

"The point remains that in the future, the survival of the livestock industry in this country depends upon the continued success of screwworm control," writes Richardson, who is a rancher as well as a scientist, "and we know of a number of problems which could arise in far less than ten years which could destroy the effectiveness of the program. The sense of urgency is clear. The need for close collaboration of all concerned—from the campus to the Congress—will continue in order to wage a successful fight against this pest."

11

THE CALIFORNIA RED SCALE AND ITS PUZZLING PARASITES

PARASITIC INSECTS ARE EXQUISITELY ATTUNED TO THE PHYSIOL-
ogy and life cycles of their hosts. No genus so clearly demon-
strates this feat of adaptation as the minute wasp *Aphytis*.

Entomologists have now described about 90 distinct
species in this extremely confusing genus. The yellowish or
grayish adults are sometimes called "golden chalcids." According to Paul DeBach and David Rosen, who wrote the de-
finitive monograph on *Aphytis*, the genus was named in
1900 by L. O. Howard for a locality in ancient Greece, presumably "because of the proximity to the Chalcidice, an area

in Northern Greece from which the name of the superfamily Chalcidoidea was derived."

Aphytis is superbly equipped for its role as a tactical weapon in biological control. It is one of those parasites that entomologists call "a good searcher," indefatigable in hunting down its hosts and parasitizing a large percentage of them. It seems to be almost free of hyperparasites in those areas to which it has been introduced, so that its population is not likely to be devastated by a serious natural enemy of its own. Furthermore, like many other efficient parasites, its life cycle is much shorter than its host's, thus enabling it to produce several generations to its host's one, and to keep up with explosive pest populations.

These wasps are the most effective natural enemies of the armored scale insects, which are distinguished by their hard shield, or covering, as opposed to softer-bodied relatives. Many of these scales live entirely free beneath the shield, usually being attached to it only during the molt. Although the *Aphytis* adult female is only about one millimeter long, she finds the shield no deterrent when she has selected an armored scale for a host. She is an ectoparasite, laying her eggs on the surface of the scale's body, rather than internally, after piercing the shield with her ovipositor. (She will not oviposit on a scale that is attached to its shield.) The one or more eggs *Aphytis* lays are fastened to the scale's body by a small adhesive pad.

Like the scale, then, the immature *Aphytis* lives under the shelter of the shield, sucking fluids through the body wall of its host. When it is fully grown, the parasite pupates beneath the shield left by the late proprietor. After emerging from the pupa, the adult chews its way to freedom through the shield or slips out from under it.

Aphytis compounds its destructiveness by doubling as a predator, the adult female often consuming the fluids of unparasitized scales that she has stabbed with her ovipositor.

The University of California at Riverside is built on the outskirts of the city, its hilly, deftly landscaped campus shielded from the southern California sun by thick stands of palms, pines, and eucalyptus trees. Riverside itself was once the center of a thriving citrus industry largely controlled by gentlemen farmers in Riverside and Orange counties, but the area now is in danger of being overwhelmed by Los Angeles' remorseless expansion. Much of the citrus industry has emigrated to the San Joaquin Valley, from Bakersfield north to Fresno, thereby complicating the longest campaign against a single insect in the history of biological control.

"The California red scale became the biggest threat to the state's orange groves after the cottony-cushion scale was controlled," said Robert F. Luck, Associate Professor of Entomology in the University's Division of Biological Control. "It is said to cost growers about 15 million dollars a year in reduced crops and spraying programs, but wherever the groves are under good biological control, the scale is almost impossible to find. Our job is to try to get that kind of control in the most important area of all—the San Joaquin Valley."

Bob Luck, a Canadian by birth, is a tall, dark-haired man, with glasses and a neatly trimmed beard. He grew up in California and, while an undergraduate at Berkeley, became interested in ecology. "During an entomology course I found out about parasitic wasps and I became hooked," he said. He went on to earn his doctorate at Berkeley and joined the biological control staff at Riverside in 1972. He has worked on the natural enemies of the Nantucket pine tip moth, which is a pest in Christmas tree plantations and nursery stock, and collected parasites of the elm leaf beetle in Spain and Morocco, but the California red scale now claims much of his attention.

His place of work is the insectary on the Riverside campus. It is a low white building, not much to look at, its windows shaded by heavy louvers to regulate light and heat. Inside, basic work has been carried on by Paul DeBach and his colleagues to unravel the mysteries that hobbled the California red scale program for most of this century. Bob Luck is part of that program.

"The stumbling block right now is that the farther the citrus groves are from the coast, the less effective the parasites become," Luck said. "The red scale is by no means the number one pest in the San Joaquin Valley. It seldom kills a tree or ruins the fruit, though it may cause twig death or leaf drop. No one has ever proved this decreases a tree's yield. But the scale does create what we call a cosmetic problem. It encrusts and sometimes blotches the skin of the fruit and makes it less attractive for the market. Even when the scale is hardly noticeable on the fruit, Japan rejects it because it doesn't want to import any scale at all—although Japan has its own infestations of the California red scale! This threat scares our growers who ship to Japan. So they use pesticides in the San Joaquin Valley and destroy the potential control to be had from the scale's natural enemies."

The California red scale (*Aonidiella aurantii*), which is named for its reddish cast, invaded the state from the Far East and was recognized as a serious pest as early as the 1870s. When Albert Koebele went to Australia on his cottony-cushion scale mission in 1889, he collected several predatory beetles that showed promise as natural enemies of the red scale. Although they managed to establish themselves in this country, they did not seriously threaten the scale.

A more effective control, in fact, came about by accident. Around 1900, a parasitic wasp, *Aphytis chrysomphali,*
appeared in California, no one is quite certain how, and
proved to be a fairly efficient natural enemy of the red scale.
This species of *Aphytis,* in turn, ran into considerable grief
at the hands of honeydew-loving ants and the measures
taken by the growers in an attempt to eradicate the red
scale. At first they used sprays consisting of soaps and oils.
Cyanide fumigation was tried around World War I, but the
scale eventually developed resistance to it. According to De-
Bach and other biological control experts, *Aphytis* was never
given an opportunity to prove its worth. Instead, by a twist of
fate, its fortuitous appearance on the California scene helped
to slow down the introduction of the red scale's most virulent foes.

During the next several decades confusion abounded.
The various species of *Aphytis* can be distinguished from
one another only by the most careful microscopic examina-
tion, the subtleties of which were apparently beyond the appreciation of early collectors and their local collaborators in
remote places abroad. Time and again species of *Aphytis*
were collected from the California red scale in China, but
then discarded because collectors thought they were dealing
with *Aphytis chrysomphali,* which they knew was already
established in California.

Similar delays in effective control came about when the
California red scale itself was confused with other small
scales, some of them not even of the same genus. In one
case, collectors sent home parasites they had taken from a
host wrongly identified as the California red scale; naturally,
the parasites were totally ineffective once they arrived in the
United States. During the middle 1930s, entomologists believed the California red scale belonged to a genus of scales
native to South America and collectors were sent there in a

fruitless search for parasites. After the mistake was discovered, collectors returned to China and located two endoparasites, or internal parasites, which, combined with *Aphytis
chrysomphali,* provided reasonably good biological control
over the scale in the citrus districts of coastal southern California.

Orange growers, however, were not satisfied. Prodded
by shippers who were becoming increasingly finicky about
even small blemishes on the skin of oranges and other citrus
fruits, the growers began turning to the latest chemical insecticides. These chemicals, in turn, created more problems
than they solved, destroying the complex assemblage of
predators and parasites in citrus groves and allowing formerly innocuous insects to escape natural controls and ex-
plode to pest proportions. Even the cottony-cushion scale,
which had faded into obscurity, enjoyed a renaissance when
DDT and other chemicals decimated the vedalia beetle; new
releases of those tireless predators quickly put the lid back
on the scale when spraying was discontinued or modified.
Yet growers, after 50 years of confusion among entomologists, were convinced that biological control was not going to
solve their problem with the California red scale. The chief
credit for researchers finally getting a grip on the problem
must go to Paul DeBach, who first learned about the art of
using natural enemies while he was at the University of California at Los Angeles during the 1930s. He was a sophomore
at the time, taking a course in economic entomology.

"Professor Harry S. Smith, the late world authority on
biological control, was invited over from the then University
of California's Citrus Experiment Station at Riverside to give
a series of lectures on biological control," DeBach has written. "This was the turning point in my professional career.
Biological control was so intellectually satisfying, so biologic-
ally intriguing and so ecologically rational a means of pest
control that I immediately opted to become a specialist in

this field and to do my Ph.D. research when the time came under Professor Smith at Riverside."

DeBach eventually became Professor of Biological Control at Riverside, where his teaching and research greatly influenced the course of this science. He also has traveled all over the world to collect natural enemies of California's most important agricultural pests. Having demonstrated their worth in his own state, he has shipped these parasites and predators abroad once more to control infestations in Greece, Italy, South Africa, Japan, and dozens of other countries.

"The modern foreign explorers can often accomplish in days or weeks what previously would have required months or a year or more," he wrote in *Biological Control by Natural Enemies,* in 1974. "Their conditions of travel and living in foreign countries are usually considerably more pleasant and health risks fewer but off-the-beaten track conditions are still grim in many places. In my own experience I have had to have armed guards in order to travel and collect in certain localities in Burma; I have lived in hovels on stilts where the toilet was a hole in the floor; have had pig-intestine soup for what I later recalled was Thanksgiving Day dinner; have had to have local permission and blessing to enter and collect in bandit-infested parts of the semi-autonomous Tribal Territories of West Pakistan and once enroute received many required penicillin shots from nearly anyone I could find, including airline stewardesses, pharmacists and medical orderlies. On the other hand, I have had some lovely collecting in Rio, Tahiti, Japan and other places."

Among his many achievements, DeBach became the world's foremost authority on *Aphytis,* doing much to untangle the taxonomy of that genus. The first real break-through occurred in 1948, with the establishment in California of *Aphytis lingnanensis*. This species had been known since early in the century as a common parasite of the California red scale in China, but no one bothered to import it to the United States because it was thought to be *Aphytis chrysomphali*. Yet once it was reared in large numbers and released in California citrus groves, it proved to be a deadlier parasite than its relative. In fact, as the newcomer flourished and spread throughout coastal areas, *Aphytis chrysomphali* virtually disappeared, hanging on only in a few coastal enclaves. In parts of the interior, however, where the greater extremes of heat and cold reduced the effectiveness of *Aphytis lingnanensis,* gaps remained in the control of the scale. During this period, DeBach and others at Riverside performed detailed experiments, using exclusion cages on citrus trees, that demonstrated the success of both species of *Aphytis* where the climate was suitable. They also attempted to develop strains of the newcomer that might be better adapted to climatic extremes.

Meanwhile, DeBach's increasing familiarity with the genus convinced him that somewhere there must be one or more species that had adapted to the harsher portions of the durable red scale's range in the Asian interior. He found one in both India and Pakistan—another of those parasites that had been rejected earlier as redundant—and named it *Aphytis melinus*. This species has been the most effective enemy of the California red scale to date. Introduced in 1957, it competed so successfully with its relatives that it extirpated *Aphytis chrysomphali* in its remaining strongholds inland and confined *Aphytis lingnanensis* to a narrow band along the coast. Entomologists had now been able to observe in detail the intense competition among three closely related species of parasites for a single host, with one parasite displacing another, and yet at each succession gaining more control. Wherever *Aphytis melinus* became dominant in southern California and was not affected by chemicals, the California red scale ceased to be a pest. But trouble loomed in the San Joaquin Valley.

DeBach, in uncertain health, retired officially from Riverside in the late 1970s, although he kept an office at the South Coast Experiment Station and continued to produce a stream of scientific papers from his hideaway in Baja California. His younger colleagues, Mike Rose and Bob Luck among them, carried on the investigations of an ecological problem that is seen to be more complex at every stage. The story is being put together bit by bit, built on hours of observation in the field or through a microscope, detailed behavioral studies, and scanning video tapes taken of the insects during their adult stages as they search for hosts. There were basic ecological questions to be answered. For instance, why had *Aphytis melinus* displaced what had appeared to be another effective parasite, *Aphytis lingnanensis?*

"The simplest way to answer that question is that *melinus* gets there first," Bob Luck explained. "*Aphytis* has what is called a haploid-diploid reproductive system. When the female in any of these species lays an unfertilized egg, the egg produces a male offspring that is haploid—it has only one set of chromosomes. If she fertilizes the egg, it produces a female that is diploid—with two sets of chromosomes. We don't understand what cues she decides to use, but she seems to be able to manipulate the sex ratio of her offspring by fertilizing the eggs. Whatever the mechanism here, she allocates the male eggs to smaller hosts, and females, or fertilized, eggs to larger hosts.

"But now we've discovered that there is a slight difference in our two species of *Aphytis*. *Melinus* utilizes a smaller-sized scale to produce a female than *lingnanensis* does. When she parasitizes a host, she paralyzes it and prevents it from developing any further. She has forestalled *lingnanensis,* she has appropriated the resource before it reaches the size her rival uses. And so there are a lot more female *melinus* being produced out there, and they are laying a lot more eggs and swamping *lingnanensis.*"

Moreover, laboratory experiments showed that, in a very warm climate, *lingnanensis* again failed to produce as many females as *melinus* did. On such observations of minute ecological shifts, Luck and his fellow workers are theorizing about what has happened as more of the citrus industry retreats to the San Joaquin Valley. This region is ringed by mountains and thus cut off from the moderating influence of the sea. The weather is consistently hot in summer, with perhaps 30 or 40 days over 100° F. There are also many days of comparative cold in winter, the temperature dropping into the 40s. The winter weather is hard on parasites, which need many days above 60° F to complete their development.

"*Aphytis* and the California red scale are essentially adapted to more of a subtropical system," Luck said. "Under optimal conditions, *Aphytis* has attuned itself to the scale, having a generation time of 15 to 20 days, while that of its host is about 45 days. But *Aphytis* parasitizes only certain stages in the life cycle of the scale. Both insects can survive in the San Joaquin Valley all right, but what happens is that the scale just sort of shuts down for a while during winter—not really in diapause, but more of a quiescent stage when it does not reproduce. So, even though there may be a large scale population present, it does not present each generation of *Aphytis* with the continuous supply of suitable stages that it needs."

The researchers have discovered a subtler reason for the problems faced by *Aphytis melinus* in the San Joaquin Valley, somewhat akin to those affecting *Aphytis lingnanensis* elsewhere. Even when functioning in the valley during cold weather, *melinus* cannot produce a high ratio of females to males. This drawback is a consequence of various elements interacting in the specific environment that is the San Joaquin Valley.

The California red scale lives almost anywhere on a tree above ground level. It grows largest, however, on the softer portions of the tree, such as new twigs, leaves, and especially fruit. The scales found on the older, woody portions of the tree are generally smaller than those on the softer parts.

Aphytis melinus, as we have seen, allocates its unfertilized eggs that will produce males to the smaller stages of its hosts and its fertilized eggs to the larger stages. But fruit, on which the scales generally grow largest, is usually picked in the winter in the San Joaquin Valley. *Melinus* is then mainly restricted, at a peak period of its reproductive cycle, to parasitizing the smaller scales found on the woody portions of the tree and thus tends to produce more males than it does in the warmer parts of its range. With fewer egg-laying females in the field, the parasite obviously is not performing at the high level of control demanded by the growers.

Fortunately, basic research in this case grades into applied science. Bob Luck spends much of his time now visiting the rearing rooms in the insectary at Riverside, noting the growth of the insects needed to overcome what used to seem to be intractable problems in the San Joaquin Valley. A few years ago it was thought that the discovery of a new species of *Aphytis* in some isolated Asian district, similar to the climatic extremes of the San Joaquin Valley, might be all that was needed to complete the campaign against the red scale. But intensive studies have shown that the bottleneck in winter is not so much the susceptibility of *Aphytis* to the climate, but rather the shortage of the resource—a continuous supply of hosts of a suitable size—it needs for success.

"We are entering a new stage," Luck said in one of the insectary rooms. "The best way to bring the populations of *Aphytis* back quickly after the winter decline is through inoculative releases. That means rearing parasites in the insectary each year and then getting them into the field. Our behavioral studies have showed us how we can produce a high ratio of females—by exposing the ovipositing parasites to the proper sizes of the host. There is still a great deal we need to know about these insects: What cues the parasite uses in searching out a host, and how she perceives its quality."

Around him, as he spoke, were small cages stocked with potatoes and lemons that were covered by thousands of tiny insects. A dim light filtered in through the louvers, while machines and ducts kept the temperature at about 74° F and the humidity at 40 to 50 percent. The low room, about seven feet high, helped to reduce the cost of air conditioning (and allowed medium-sized entomologists to pluck escaped insects from the ceiling).

"These are scales in the crawler stage," Luck explained. "They'll take about 30 days to develop into adults. We collect them in the field and bring them in here in plastic cups. We wash the potatoes to make sure there are no other insects on them, lay the scales out on those trays in the cages, and allow them to settle on potatoes. When the scales mature and begin producing crawlers, we use them to infest new potatoes."

Every item in the room is geared for the efficient rearing of large numbers of scales. Because lemons or other citrus fruit would quickly dry up under the assault of so many tiny sucking mouths, the crawlers are grown on potatoes.

The scale culture requires constant attention. If, for instance, mealybugs got into the potatoes they would produce honeydew and create what Luck described as "an awful mess." To guard against that possibility, the researchers maintain, at low levels, a colony of ladybird beetles (genus *Cryptolaemus*) that attack only mealybugs.

"It's a *real* calamity if predatory mites get in here," Luck said. "We might have to put the entire cage in a hot box and heat it to 170° F to kill everything. If the whole room got infested, we'd have to destroy the entire culture and fumigate, then go out in the field and get more red scales to start a new culture."

Eventually, when the scale has reached the proper stage, the infested fruit is transferred to a room where the breeding stock of *Aphytis* is held. The female parasites are turned loose on the scale. After about 12 days, when a new generation of parasites emerge, researchers prepare them for life in the wild.

"We inject CO_2 and ether into the cabinet, which anesthetizes the parasites within two minutes," Luck said. "We open the cabinet and push down those levers that you see on the side. We slide out the floor of the cabinet with the anesthetized parasites and brush about 3,000 of them through a funnel into each container, which is a half-pint carton. We pack the parasites in dry excelsior, which has been soaked in honey and water, and take them to the field for release.

In this case, "the field" is a citrus grove in need of an inoculative release of *Aphytis melinus*. The entomologists work in nine-tree blocks, releasing the parasites into a tree that is adjacent to eight others. From then on, it is up to *Aphytis*.

"There is a whole complex of pests on citrus trees in the San Joaquin Valley, and the red scale is just one of them," Bob Luck said. "But they all have natural enemies.

Our experience indicates that with a sound biological control program we can cut the grower's costs while preserving the quality of the fruit. There are added benefits—like reducing hazards for farm workers. In the long run, however, our toughest job will be to re-educate most of the growers."

12

THE ALFALFA LEAF MINER
The French Connection

THE ALFALFA BLOTCH LEAF MINER (*AGROMYZA FRONTELLA*) IS one of those insects that makes us more aware than ever of the dramatic difference in size that helps to distinguish their view of reality from ours. The last part of its vernacular name tells the story: It refers to that critical period when this insect lives entirely between the walls formed by the upper and lower surfaces of an alfalfa leaflet.

This leaf miner is a native of Europe, where its population is low and poses no menace to farmers. It was first reported in the United States in 1969, although it probably invaded some years earlier. Spreading rapidly through the northeastern states and into Canada, it became a pest on alfalfa, which is North America's most important livestock forage crop. Often 70 or more of these insects infested the leaflets on a single stem.

Representatives of several insect orders, including beetles, wasps, and moths, have adopted leaf mining as a way of life for their larval stage. The alfalfa blotch leaf miner is one of the Diptera. The adult, a small, dull-black, humpbacked fly, emerges in spring from a hiding place on the ground where it has spent the winter as a pupa. It feeds on alfalfa, puncturing the leaflets with its ovipositor and then turning around to suck the plant juices. The fly also lays from one to three eggs in a leaflet, and when they are abundant, several flies may lay as many as 20 eggs in a single leaflet.

The eggs hatch into yellowish specks that are the leaf miner maggots. The maggots, or larvae, live among, and feed on, the nourishing plant cells under the leaflets' "skin." As these blind, legless maggots make their way through the plant cells, they push apart the upper and lower surfaces of the leaflets. Air slides in between the surfaces, giving the leaflets a grayish-white cast that is referred to as a "blotch." A severe infestation robs the plant of its nutritive value and may even cause some of its mined leaves to wilt and fall.

R. M. Hendrickson, Jr., and S. E. Barth of the staff of the USDA's Beneficial Insects Research Laboratory in Newark, Delaware, studied the ecology of the leaf miner infestation. Two other closely related leaf miners, native to North America, feed on alfalfa, yet are kept under such tight control by a battery of natural enemies that they remain as unobtrusive in the United States and Canada as the alfalfa blotch leaf miner does in Europe. The entomologists discovered that at least 14 species of native parasites were also attacking the foreign leaf miner within a few years after it crossed the Atlantic.

Why, then, was the new leaf miner flourishing? Hendrickson and Barth conducted a series of experiments, bringing infested alfalfa to the laboratory, studying the behavior of the three leaf miner species, and observing their interaction with their parasites. They documented the fact that the native parasites were perfectly synchronized with the native leaf miners. However, they observed that there was very little parasitism in spring on the pupal stage of the alfalfa blotch leaf miner. The parasites appeared in the spring two or three weeks after the adult flies emerged, so they were too late to be effective against the first generation of leaf miners and there were too few present to control the second generation. Later in the season, the native parasites proved much more effective against the foreign pest.

The USDA had already established the apparatus for responding to this emergency.

"I was working on the alfalfa weevil with the Illinois Natural History Survey during the middle 1960s when I heard about the opening at the Paris laboratory," Richard J. Dysart said. "I applied and was given the position, so I had my first experience living in a metropolis. I lived in the Sixteenth Arrondisement, near the Porte St. Cloud, 30 or 40 minutes from the laboratory. I never quite got used to the traffic jams—they were fierce—but I never regretted applying for the position."

Dysart is now in charge of the USDA's Delaware Beneficial Insects Research Laboratory, which serves as the terminal for most shipments of insects to the United States from abroad. Here, natural enemies of all kinds are received, opened, examined with extraordinary care, and then sent on to other points for release against serious pests. But Dysart used to be on the other end of the line, collecting insects all over Europe and screening them for shipment across the Atlantic.

"I had had some French in school, but I couldn't speak the language when I arrived in Paris," Dysart continued. "In France, you know, they're not eager to share with you the 10 or 12 years of English they've had in school, so you have to learn to speak *their* language. I enrolled at the Alliance Française in Paris for a crash course. It was a strait-laced and old fashioned place where everything had to be said in French, but it was the best way to learn. It also helped that some of the other scientists and staff members at the USDA laboratory were French, so we spoke it there all the time, too."

Dysart collected in Europe for the six customary years at a critical time in the history of the USDA's European Parasite Laboratory. It was in the immediate aftermath of *Silent Spring,* when there was a new urgency in the USDA to give at least the appearance of looking for alternatives to chemical control. The search for natural enemies of the cereal leaf beetle and the alfalfa weevil was in high gear and the agency was about to revive its parasite program against the gypsy moth. If the European Parasite Laboratory has been a neglected arm of USDA's vast bureaucratic empire, given short shrift in both operating funds and publicity (and for most of its history existing more like a floating crap game than a respected institution), it has nonetheless compiled an impressive record of service. The University of California's Division of Biological Control, as we have seen, maintains a revolving fund that allows its collectors to leave for foreign parts almost at a moment's notice, since the USDA's creaking budgetary machinery often throws up obstacles to a hasty departure. The European laboratory's existence at least gives the agency a collector near the scene of action when exotic natural enemies are badly needed.

The European Parasite Laboratory owes its creation to the corn borer, which invaded the United States during World War I. "In 1919 L. O. Howard, the chief of the Bureau of Entomology of the USDA, engaged W. R. Thompson to produce natural enemies of the corn borer in Europe," writes John J. Drea of the USDA in a brief historical sketch of the laboratory. "Young Thompson was well qualified for the task at hand. Prior to World War I, he had searched for parasites of the alfalfa weevil in Italy. Furthermore, he had just completed his doctoral studies in Paris and was well acquainted with this country and its language."

Thompson rented a villa in the corn-growing region of southwestern France and established a USDA laboratory there. During the next 20 years, the laboratory moved from place to place in France under Thompson and his successor, H. L. Parker, the exact location depending on its proximity to the haunts of promising natural enemies. According to Drea, by 1931, the laboratory was employing 430 temporary field workers to collect corn borer larvae for shipment to the United States. Between 2 and 3 million parasites of more than a dozen species emerged from the host larvae. Only one of them (a parasite named for W. R. Thompson) turned out to be of any value, but it helped to restrain the borer until resistant hybrid varieties of corn became available to farmers. Meanwhile, like almost everything else in France, the laboratory gravitated toward Paris, and the USDA rented a house in suburban St. Cloud.

"Soon the lack of space became a problem due to mass collection of plant material, especially for studies of the Hessian fly," Drea writes. "Therefore, in 1939 an insectary was built in the neighboring town of Rueil-Malmaison. No sooner was the insectary completed than World War II put an end to all activities for the duration of the war. The European Parasite Laboratory was closed down, the American personnel were transferred to Montevideo, Uruguay, the local personnel were released, and the equipment and collections were placed in the cellar of the American Embassy in Paris. Throughout the war years, J. Balakine, a local employee of the laboratory, checked the collection and equipment each week. The only loss during this trying period was his own bicycle, confiscated by the occupying troops."

The laboratory resumed its nomadic existence after World War II, although within the suburbs of Paris. When Dick Dysart arrived in 1965, the laboratory was in a renovated stable in Nanterre; from there the staff moved to an old abbey in Gif-sur-Yvette, and later still to a house in Sèvres. Dysart replaced Reece Sailer as the facility's director in 1966. It has grown in recent years to employ a staff of eleven, including entomologists, technicians, a secretary, and a gardener-custodian.

"After I was there about nine months I began going out on field trips with French technicians, and by that time I could speak with them comfortably," Dysart said. "I did most of my collecting for parasites of the cereal leaf beetle and the alfalfa weevil. We went on most of our trips by car even as far as Sweden. When we were looking for the cereal leaf beetle parasites, we would keep alert for fields along the road where oats were growing. If we saw that it was at the right stage of growth for the pest to be active, we'd take a closer look to see if there was any damage, and if so we'd know there was a good chance of coming up with the host, at least.

"We would pull over to the side and jump out with our nets. Sometimes we'd encounter a farmer who was angry about finding us tramping around in his field, but by and large we had no problem. If we had stopped to ask permission to go on his land, we would have wasted a lot of time going to the farmhouse and hunting him up. So we would just take our 100 sweeps with the net, and then clear out."

The laboratory received many of its parasites from Communist countries. Through a federal program called Public Law-480, there was money available to pay for acquiring insects from foreign collaborators, and Dysart kept in touch with collectors in Poland and Yugoslavia. He also traveled to the Soviet Union in search of cereal leaf beetle parasites.

"Any American representing the government is looked on with suspicion over there," he said. "I was there for two months. The problem was that I couldn't travel as freely as in Western Europe and I had to make arrangements in advance to visit the big state farms. But I had an interpreter with me at all times who was furnished by the Ministry of Agriculture, and that helped in ironing out a lot of problems. The cereal leaf beetle is not usually a serious pest in the Soviet Union, though I was fortunate enough to see damaging outbreaks in several fields—not large acreage but a high density of beetles. So I did some collecting there, mainly to see if the complex of parasites was similar to that in Western Europe and to add to our gene pool for the United States.

"I also did some of what we call 'sidewalk collecting,' wandering around in cities and looking in parks. Sometimes the Russians had little demonstration gardens. I would walk in and pick up some material and just stick it in my pocket."

One of the objectives of Dysart's auto trip to Sweden was to collect parasites of the alfalfa weevil, which was then a major pest in the northeastern United States. A parasite that emerged from some of the hosts collected in southern Sweden was the ichneumon wasp *Bathyplectes stenostigma*.

"We shipped the parasite back to the United States, where it was released in the East and Midwest," Dysart said. "We had high hopes for it there, especially in the Northeast, because we had taken it from a low, wet region along the Baltic Sea where the climate was similar to that of our northeastern states. Yet it never became established there. Curiously, it was also released in the Rocky Mountain States—in a high, dry region—and it has done very well! That's why we go on collecting and experimenting. You can never predict where a parasite will fit in."

Much of Dysart's time was spent in the laboratory outside Paris, screening specimens he had collected himself or received from European collaborators. The work was never easy in the series of rented buildings that housed the facility, for upstairs bedrooms had to make do as research labs and whatever foliage was needed to maintain live collections came from a small garden in the backyard. Sometimes he shipped material to the United States in the form of pupal cells containing parasites, sometimes as adult parasites that had already emerged from their hosts.

"We had excellent air freight connections to Philadelphia, much better than we have now," he said. "Late in the 1960s, Air France flew freight non-stop from Paris to Philadelphia. Now they fly the freight to Frankfurt and it comes from there to Philadelphia on Lufthansa. That's two days minimum. Sending adults can be risky with that extra day, especially when the airlines' schedules are constantly changing. Maybe they had the right idea in the old days. Robert van den Bosch and some of those other collectors in the years after World War II used to go to the bars where airline crews hung out. They would give them the packages, and one of the crew would call the quarantine laboratory as soon as they landed in the States."

During the European laboratory's early years, there was no general importation facility to receive its shipments to the United States. The USDA set up a small quarantine laboratory in Hoboken, New Jersey, to receive parasites of the corn borer; later, there was a laboratory in Melrose Park, Massachusetts, to handle parasites of the gypsy moth and brown-tail moth, and one in Moorestown, New Jersey, for those of the Japanese beetle and Oriental fruit moth. The latter eventually became the central quarantine station for USDA imports and Dysart joined the laboratory when he returned from Europe in 1971. After this facility moved to Delaware a few years later, Dysart moved with it. He is now its director and one of the five scientists assigned permanently to the laboratory's research and quarantine staffs.

Shipments arrive from abroad once or twice a week. Some of the material remains in the laboratory for research or later release by members of the staff. The bulk of the material, however, comes in chiefly to pass through quarantine, after which the staff repackages it and sends it on to scientists in the U. S. Forest Service, state agriculture departments, or various universities for projects of their own. (There is also a small quarantine station in Gainesville, Florida, which receives beneficial insects collected, for the most part, in the tropics.)

"We have to be very careful when screening the material to see that it doesn't contain anything we don't want in this country, as for instance a plant-feeding pest or a hyperparasite," Dysart explained. "Sometimes these shipments are large and they can contain a diversity of insects. If the foreign collector comes from some obscure country, we can get some pretty weird stuff—a lot of surprises. But if the package comes from the Paris laboratory we generally know in advance what's in there."

The USDA also maintains an Asian laboratory, staffed by one scientist and a technician. In the beginning it was located in Japan, but the agency preferred closer contact with a great variety of insects on the mainland and the laboratory recently moved to South Korea.

"China has a widespread population of gypsy moths and we're all very much interested in the complex of parasites to be found there," Dysart said. "Not long ago we were able to send a party of entomologists into China— Paul Schaefer, who had opened our Asian laboratory, Ron Weseloh of Connecticut, and some others. Schaefer planned to bring back live insects, but at the last minute the Chinese forestry officials wouldn't let him keep the material. It was quite a shock. But I can understand why they are wary of Western scientists. It used to be that any time a Western entomologist found a new insect in Asia he would describe it and name it and then put it in a museum in London or Paris or Washington, anywhere but Peking. But I think that gradually we'll be able to work something out with the Chinese."

When Dysart led his visitor into the quarantine laboratory it was clear that other pipelines from Asia had remained open. The laboratory is carefully sealed off from the rest of the building by a series of heavy doors along a zigzag hallway and is serviced by separate heating and air conditioning systems. Dysart held up a container of parasites that had arrived the day before from India. It was a small thermos bottle of Indian manufacture, painted all over with decorative scenes of gardens and gamboling children of the sort that might have illustrated a sentimental book 100 years ago.

"It's a clever way to ship these very minute insects," Dysart said admiringly. "Seven days enroute, and they all made it. They were shipped by one of our collaborators with the Commonwealth Institute in India. Those people are very capable. We don't get any surprises from them."

Nevertheless, every package is handled as if it might contain the nastiest of surprises. A shipment is always picked up at the Philadelphia International Airport by a staff member immediately after it has been released by a plant pest quarantine officer. The staff member brings it to one of the four "high security cubicles" in the quarantine laboratory. Only natural light enters the cubicle; the window faces north to keep out all direct rays of the sun. The cubicle is kept uncomfortably cool so that if an insect pops unexpectedly out of the package (which is opened inside a sleeve-cage next to the window) it will remain comparatively inactive. If it can fly at all, it darts toward the light coming through the window and is trapped in a transparent box, where it can be easily recaptured. The cubicle itself is stocked with all the implements, vials, and other equipment staff members might require so that they do not have to go outside quarantine until they have finished the screening process.

"When we know what is in the package we bring it into a more comfortable room for a closer look," Dysart said. "If we find that other hosts or hyperparasites have somehow slipped in with the shipment we kill them right away. Many of the shipments can be cleared quickly and go out from here the same day. The package on that table over there just came in from Poland and we're shipping it on to a scientist in Oregon who requested the material. But sometimes we may have to hold the material here and rear it to find out more about it."

As Dysart walked through the laboratory, he commented on the various insects and collections. A large refrigerator held diapausing parasites of lygus bugs that would remain there through the winter until spring; some of them would then be released in New Jersey and the rest sent on to a laboratory near Tucson for release in another control program. A cage held a species of small parasitic wasps from Japan that had been under observation at Newark for more than two years. Staff members, aware that this species attacked certain Coccinellids, were concerned that it might parasitize beneficial ladybird beetles in the United States. Farther along, a technician was observing a cage that held what appeared to be half a dozen species of ladybird beetles, some brightly spotted, others dark without markings. A museum scientist had originally identified them as males and females of separate species, but the staff at Newark has since discovered they are all members of a single, polymorphic, aphid-eating species.

"We keep a culture of gypsy moths that we've brought in from the field to see if they are parasitized," Dysart said, pointing to another cage. "These wild larvae are sometimes difficult to rear because if they have been feeding on oak they tend to eat nothing but oak. So we have to try to transfer them to an artificial diet which is made up and sold by a firm in New Jersey. The gypsy moth is the only insect in the laboratory now that accepts an artificial diet. We are also rearing Mexican bean beetles, and that project requires a lot of space because we have to grow bean plants for them."

One of the staff scientists in the laboratory that day was Robert M. Hendrickson, Jr., whose career began in a burst of glory. He is another in that legion of young biological control specialists who studied at the University of California at Riverside with Paul DeBach. When he joined the staff of the Newark laboratory in 1974 he was immediately assigned to work on a rather alarming new pest, the alfalfa blotch leaf miner.

"This pest arrived in the Northeast from Europe just at a time when the biological control people were trying to persuade farmers to cut down their spraying for the alfalfa weevil," Hendrickson said. "The leaf miner doesn't do a lot of damage, and by the time you notice it's even in the alfalfa field most of whatever damage there is has already been done. So it's really a waste of money to spray. But the insect turns an alfalfa field a light brownish-gray and it looks awful, and the farmers really get upset about it. They *know* something is wrong and they want to hit it with chemicals right away."

The sudden appearance of the leaf miner posed a double threat to what was already a well-established parasite program for alfalfa weevils. If the growers resumed spraying, they would wipe out the parasites the USDA had put in the fields. There was a more subtle threat, this one affecting the plan to collect successful weevil parasites from fields where they had become established in the Northeast and ship them to the Midwest in an attempt to increase their range. But the plan might have to be abandoned: Biological control specialists ordinarily may ship parasitized alfalfa weevils almost anywhere, because these pests are established almost everywhere; under quarantine regulations, however, they could not ship any material from northeastern alfalfa fields to other parts of the country because they might inadvertently increase the leaf miner's range as well.

And so a call went to the European Parasite Laboratory to scour the continent for natural enemies of the alfalfa blotch leaf miner. John Drea of the laboratory's staff set out with a French technician, Denis Jeandel, to see what they could turn up.

"Jeandel was a real eager beaver," Hendrickson said. "He would spend all day looking for the host. He covered alfalfa fields in France, Denmark, West Germany, Austria, Switzerland, Luxembourg—and even Liechtenstein! He sent us *thousands* of hosts. We reared 14 species of parasites from these hosts, most of them wasps, but also Diptera. We decided to take all of them for release and let nature decide.

Ironically, the parasite that Drea found most abundant on the leaf miner in Europe didn't make it here at all."

Hendrickson and S. E. Barth, who assisted him, already had an idea which of the introduced parasites would be most helpful to them. They knew from their surveys that less then 2 percent of the leaf miner's pupae in the United States were parasitized by native natural enemies; they had found that the pupae sent to them from Europe were parasitized at a rate of better than 25 percent. "It appeared that the ecological niche of a parasite emerging from the host puparium was nearly empty in the United States," Hendrickson said.

Hendrickson and Barth concentrated on two wasps that were endoparasites of leaf miner larvae, which emerge from the pupae, and reared them in the laboratory at Newark. They confined adult parasites in cages with alfalfa plants infested by leaf miner larvae. After the parasites laid their eggs in the hosts, the entomologists set the plants on their sides so that the mature parasitized hosts dropped from the leaflets into trays of moist vermiculite. When the adult parasites emerged after the hosts pupated, Hendrickson and Barth collected them for release in alfalfa fields.

By the early 1980s, the program was an astonishing success. "The parasites are now established in 15 states," Hendrickson said. "In fields where the parasites have had four or five years to build up their numbers, we are getting over 70 percent parasitism in the pests and they are really hard to find. The farmers don't even *think* of spraying now. It's really spectacular, if I do say so myself."

The campaign against the crop's real menace, the alfalfa weevil, could thus go on unimpeded.

13

THE ALFALFA WEEVIL
The Six-Pak Solution

THE ALFALFA WEEVIL (*HYPERA POSTICA*) ARRIVED IN THE United States from Europe about 1900, perhaps in the soil of imported nursery stock or the household effects of immigrants. Somehow it leapfrogged the East and Midwest to appear as a pest near Salt Lake City in 1904. It took another 40 years to establish itself in the Atlantic states, where it soon became the major threat to alfalfa. By 1970, it was established in all 48 contiguous states.

Its erratic course, as it spread into every corner of the nation, should not have surprised anyone. Like the gypsy moth, the cotton boll weevil, and other major insect pests, the alfalfa weevil is a highly opportunistic creature. The adults are accomplished long-distance flyers, and where their wings won't take them, humans will; they hitch rides in baled hay or other agricultural products, which may have been the medium they took to reach Utah early in the century. Because alfalfa is the nation's most valuable hay crop and a staple of livestock raising almost everywhere, the alfalfa weevil's omnipresence was foreordained.

The adult weevil, about three-sixteenths of an inch long, is light brown, with a broad dark stripe extending about halfway down its back; it turns dark brown or nearly black with age. It has the long snout characteristic of its family, the Curculionidae or snout beetles. The chewing mouthparts are at the tip of this snout. A part of each season's adults survive to overwinter among stubble in the fields or on their brushy borders.

These are the weevils that are active in alfalfa fields in early spring. The female chews holes in the tender parts of alfalfa stems, lays clusters of eggs (usually anywhere from two to twenty-five) in each stem, and seals the holes with fecal plugs. The lemon-yellow eggs, about one-fiftieth of an inch long and darkening with age, hatch in a week or two as a rule, although cool weather may retard their development. The larvae, when they hatch, are yellow with shiny black heads and are so small that they can hide within the unopened leaves at the tips of a plant. There they feed unceasingly, growing quickly and shedding their skins and then moving to the larger opened leaves as they approach maturity. The larvae, like the young of most other plant-feeding insects, are much more destructive than the adults.

After three or four weeks, the larvae spin cocoons on the plants or in the leaf litter below. They pupate and emerge, within one or two weeks, to mate and produce new

The alfalfa weevil (below) and its larva.

generations in milder climates, but in colder realms they simply feed for a time before proceeding into their winter quarters.

William H. Day, a research entomologist at the Beneficial Insects Research Laboratory in Newark, Delaware, is a tall, outgoing man who obviously likes what he is doing, which is field research on the parasites of the alfalfa weevil. "It was an unexpected by-product of taking on administrative duties here that I am still able to work on the alfalfa weevil," he said. In any event, Bill Day can look upon the alfalfa weevil program with an equanimity that was not possible at the time he joined the laboratory in 1965, when it was still located in Moorestown, New Jersey.

"My excitement at being assigned to the project, which had all the top people working on it, was quickly dampened," he recalled. "I went out in the fields and saw the devastation this insect was causing—it was just eating all the leaves off the plants—and I said to myself, 'Boy, I'm really with a loser. We're going to go down in flames!'"

The project was as important as it was daunting. American farmers grow between 25 and 30 million acres of alfalfa each year. This legume is extremely desirable as a livestock feed, being rich in protein, calcium, and vitamins, giving a high yield, and not least important, it is exceptionally palatable. Under normal conditions, it is a comparatively easy crop to grow. A skillful farmer in the northern states can maintain a flourishing alfalfa field for 15 or 20 years without cultivation or reseeding; aside from treating his fields with herbicides, he simply fertilizes the crop in the fall, and harvests through the following spring, summer, and fall. And all the time, this legume is building up nitrogen and organic matter in the soil.

Dairy farmers are the principal growers and users of the crop, seldom having enough left for sale, except perhaps when a higher than normal rainfall produces a bumper crop. Others grow alfalfa for race horses or for processing plants that make high protein pellets for turkeys or rabbits. When alfalfa is available in the Northeast, it may sell for anywhere from 5 to 120 dollars or more a ton. Considering large acreage, the minimal investment, and the utility of the yield, it is one of that region's most valuable crops. The alfalfa weevil robs the crop of much of its nutrition and sometimes almost destroys a field. Although estimating insect damage in dollars is an inexact and sometimes suspiciously subjective process, a modest annual estimate by the USDA charges weevils with "chewing up" 80 million dollars worth of alfalfa.

"I can remember going into alfalfa fields and with one sweep of the net collecting 300 larvae," an entomologist with the Newark laboratory has said. "That's incredible! Everything around us looked gray, and the stems crackled underfoot when we walked through the field. There was just nothing left but skeletonized leaves."

That was a common scene all through the eastern part of the United States during the 1960s, when Bill Day came to the project. At least, they were still growing the crop in the northern states. Farmers in the Southeast abandoned alfalfa, turning to various grasses and other forage crops, because of the high cost of chemicals in weevil control. Some critics felt that the USDA was little more than a research arm of the chemical industry at the time. Its highest-ranking administrators and scientists thought mainly in terms of chemicals, considering biological control to be a public relations ploy, a sop to environmentalists, and not a technique that worked in the real world. But new blood was entering the agency.

Bill Day's interest in the natural world goes back to his

childhood. His uncles were outdoorsmen and hunters, and he grew up across from a large park in Wilmington, Delaware, where he liked to watch wild things of all kinds; birds and rabbits could be seen only from a distance, he noted, whereas insects could be picked up and examined by a curious youngster. When he enrolled at the University of Delaware he drifted toward biology, although it was the teacher and not the discipline that snared him.

"There was a professor of plant pathology who was a dynamic speaker, but a student couldn't major in this subject without having some background in entomology," Day said. "I'm going to have to take entomology, too, I realized. So I studied entomology under Dale Bray and Charles Triplehorn, who were also very good lecturers, and that decided my choice of a career. I did my graduate work at Cornell under Art Rawlins. During the summers I worked on Cornell's research farm on Long Island and became interested in why aphids were so variable on potatoes from year to year. I concluded that insecticides were an important part of this alternating scarcity and glut of aphids. I was also fascinated to see the part that diseases and other natural enemies could play in controlling a pest."

The Beneficial Insects Research Laboratory was just beginning to put some order into the parasite program against the alfalfa weevil when Day joined the staff. The USDA had released several species against this pest in Utah earlier in the century, and one species anyway had certainly played a part in reducing the weevil's destructiveness throughout the West. Another parasite apparently had entered the eastern United States within the invading weevils themselves, thus becoming what specialists call a "freebie."

But during the late 1950s, and picking up momentum in the 1960s, there was a flood of parasites into the country from the Paris laboratory.

"By 1968, I knew from my surveys in limited areas that the parasites were making a difference, and we were trying to step up the number of releases," Day said. "But we weren't getting much cooperation from the state extension people. They thought these early successes were a fluke. Most of them had never seen a biological control program work, so it was understandable that they didn't believe it *could* work. There had been very few of these projects on the East Coast—only the European corn borer, which was not successful, and the Japanese beetle, and a few others which were only partly successful. Oh, yes, some of the extension researchers were convinced by the evidence in the fields, but the agents who were in frequent contact with the farmers just didn't believe in it.

"Then some of the progressive farmers stopped spraying for the weevil. A neighboring farmer would say, 'Gee, I spent 500 hundred dollars on chemicals this year. That fellow didn't spray, but he had no more trouble with the weevil than I did.' A big break occurred when the New Jersey State Department of Agriculture began to cooperate with us. They helped to distribute parasites there and became the leader among the states, indirectly benefiting New York and Pennsylvania because in three to five years the parasites built up to such high numbers that the surplus flew right out of New Jersey into nearby states."

Within an astonishingly brief period, the alfalfa weevil in the Northeast was under excellent control by the introduced parasites. By 1968, the tide had turned; during the 1970s, the farmers in that region (where spraying for the weevil had been almost mandatory) gave up chemicals so that by 1980 less than 20 percent of the acreage was being treated. ("And most of those people are simply prolonging their problems," Day said.) There was a momentary flurry of concern when, as the weevils became scarce, some of the parasites dipped sharply in numbers, but they soon recovered and now remain in equilibrium with the pest. Some farmers in the Southeast have even started growing alfalfa again.

Because the alfalfa weevil was still a threat in the Midwest and in the Rocky Mountain region, the USDA decided to follow up on its triumph and strengthen its alfalfa weevil program in those parts of the country. Coincidentally with this decision, the agency was winding down another successful campaign, that waged against the cereal leaf beetle, while the concerned states assumed more responsibilities for its final stages. The Parasite Rearing Laboratory in Niles, Michigan, was now free to accept another assignment. Centrally located and with a staff under director Tom Burger that had gained experience working with the cereal leaf beetle, the Niles laboratory was the logical place from which to coordinate the expanded campaign.

"The alfalfa weevil is really two different beasts," said Milt Holmes, a researcher on the Niles staff. "Much of our information about the weevil is based on research that was done on the eastern strain, so we've had to make adjustments in dealing with the western strain. In the East, the weevils come out in the spring, mate and oviposit, and produce a new generation of larvae. The new summer adults come out, feed for a time, and then scatter—but in fall most come *back* into the fields.

Some of them mate and lay eggs in the fall, especially in the Southeast, though most eggs laid farther north at that time are killed by cold weather. But the western strain of weevils leaves the field in midsummer, when it is very hot and dry, and holes up in woodlots or even in rock piles. They don't return to the field until the following spring. So each area has a different time when the larvae are at their peak density. We have to know for each location what is the ideal time to ship in parasitized larvae for release."

Because alfalfa is grown everywhere in the United States and the weevil is a pest wherever the crop is grown, the parasite complex is now a maze of interlocking parts. The USDA entomologists released eleven different species in the East, although only five successfully established themselves, and several in the West. With the creation of a Parasite Detection Survey, which has expanded to all parts of the country, the Niles laboratory is putting together what is almost a primer on how a smooth-functioning parasite complex deals with a widespread pest. There are parasites that attack every stage of the weevil's life cycle. Here, in brief, is a "rogues' gallery" of the alfalfa weevil's most tenacious natural enemies:

• BATHYPLECTES ANURUS (pronounced bath-uh-*pleck*-tees ah-*noor*-us). This is the "jumping parasite." The staff of the European Parasite Laboratory collected it chiefly in France (although small numbers were found later in Sicily and the Soviet Union) and shipped it to Moorestown in 1960. It builds up slowly in the field where it is released, but with an abundance of hosts and more releases, researchers eventually found it in 17 eastern states and Ontario.

Even in the strange world of parasites, this wasp has an unusual life history, combining a comparatively long life with a very brief spurt of activity. The female oviposits in young weevil larvae in spring. When the egg hatches, the parasite consumes its host, spins a cocoon inside the weevil's cocoon, in foliage, or among the leaf litter in the field, and enters diapause. In the autumn, the parasite comes out of diapause and, while still in the cocoon, pupates and changes into an adult—and immediately reenters diapause. Its two long periods of diapause are almost unique in the insect world. The parasite overwinters in this state and the female emerges from its cocoon in spring to mate and find hosts for her eggs.

Almost as unusual is the parasite's behavior while encased in its hard, brown, football-shaped cocoon. Entomologists had observed the cocoon repeatedly jump several centimeters upward in the field. In 1970, Bill Day proved it is responding to environmental conditions, its jumping rate increasing with heat or light as it escapes the direct rays of the sun after the three or four cuttings of the crops during summer. He found that the jumping also decreases hyperparasitism.

• BATHYPLECTES CURCULIONIS (kur-kew-lee-*ohn*-iss). At one time, this was probably the most successful parasite on the alfalfa weevil. Collectors sent it from Italy to Utah as early as 1911. It became established there and, when the weevil invaded the East, became a prime candidate for the control program. Brought from the West by the USDA, it "spread naturally at a phenomenal rate."

This wasp is also a larval parasite. Despite the fact that it is the only established parasite for which the weevil has a defense, sometimes encapsulating and destroying its egg, *B. curculionis* had a high rate of success until recent years. Encapsulation seems to be taking its toll, however. *B. anurus*, which does not have this handicap, has all but crowded out its close relative in the Northeast.

• BATHYPLECTES STENOSTIGMA (sten-oh-*stig*-mah). This is the wasp Dick Dysart collected in southern Sweden in the belief that it might thrive in the northeastern United States. Although it failed to live up to its promise in that region, the USDA did not give up on it. The agency released it in Colorado, where it gained a foothold and became an effective natural enemy.

After killing its host, the parasite drops to the ground and spins a cocoon, just under four millimeters long and resembling a brown paper bag, and spends the winter in diapause. Like the other members of its genus, *B. stenostigma* is vulnerable to hyperparasites because it spends so many months exposed on the ground in its cocoon.

• DIBRACHOIDES DYNASTES (dy-brah-*koy*-dees dy-*nass*-tees). This metallic-green parasite, which attacks the weevil pupa, was imported from southern France beginning in 1959. It is a rather startling experience to watch the drama being played out in a succession of scenes under a high-powered microscope in the laboratory: the weevil larva spinning its loosely woven, curiously beautiful cocoon; the female parasite mounting the cocoon to plunge its ovipositor through the wide-mesh covering into the inert but living bundle below; the emerging parasite larvae, which appear as tiny maggots to the naked eye, now magnified and feeding in clusters of three or four inside the cocoon's loose net, but on the pupa's surface, like vultures on the bloated carcass of a wildebeest.

"The parasite seems to prefer a mild climate and we haven't recovered any of them after our releases in the Northeast," Holmes said. "It overwinters as an adult and a lot of them die off, so in most cases they start with a low population in the spring. We think it will do better in the South, especially because the time frame between hosts is reduced there. In the South it will have hosts in the field much of the year, whereas in the North it has three months of hosts and nine months of nothing."

• MICROCTONUS AETHIOPOIDES (my-*croc*-tuh-nus ee-thee-oh-*poy*-dees). This wasp is one of the two most important parasites (*B. anurus* being the other) on the alfalfa weevil east of the Mississippi River. Collected in France, it was reared at the Moorestown Laboratory and released in New Jersey in 1957.

It goes through two generations a year, parasitizing the adult weevils. In spring the reddish-brown female, which is about three millimeters long, stalks the "old" overwintering adults. She doggedly follows her victim until she is able to thrust her ovipositor through the segments on the underside of the abdomen. This species may parasitize anywhere from 70 to 90 percent of the weevils.

After the young parasite hatches and kills the host, it emerges that summer to produce a new generation. The females oviposit in sexually immature weevils. The young parasites of the second generation overwinter within the adult weevils. Although they do not kill their hosts until the following spring, they render them sterile.

• MICROCTONUS COLESI (*kohl*-zy). Evolution has fitted this wasp beautifully to the life history of the alfalfa weevil. Although it has but one generation a year and is active only one month out of twelve, it is an important member of the parasite complex. After the female oviposits in a weevil larva, its offspring hatch and remain in diapause throughout the developmental stages of the host; the weevil metamorphoses from larva to pupa to adult, feeds a bit in the fall, and retires for the winter, all the while carrying the somnolent yet deadly parasite within it. The following spring, the parasite comes out of diapause and kills the weevil before it can lay its eggs.

This species is the "freebie," for it was not purposely

imported to the New World. It reproduces by parthenogenesis, and no one has ever found a male *M. colesi*. Entomologists believe that its establishment in the United States resulted from a small number of individuals that arrived inside the weevils themselves. Because the females do not have to mate, the species can establish itself in new areas with the release of comparatively few individuals.

• PATASSON LUNA (pah-*tah*-son *loo*-nah). This is another species that was imported from Italy to Utah earlier in the century. Although it was never released in the East, it appeared in New York State in the 1950s, causing entomologists to speculate that it was brought east years ago when this species was still confused with another parasite. (*Patasson luna* is very tiny, about one millimeter long, and much confusion existed about its identity before sophisticated optical instruments became common.)

Patasson luna is the egg parasite on the weevil, earning a reputation among biological control specialists as "a good searcher." It shows great sensitivity to the host by detecting the feces-covered weevil eggs hidden in alfalfa stems. Like other egg parasites, it is most effective when the host population is under stress. In that event, the host lays fewer eggs and the parasite is able to reach a larger percentage of those packed into the stem.

• TETRASTICHUS INCERTUS (teh-*trass*-tih-kus in-*ser*-tus). This wasp appeared in the United States in 1960 when six individuals emerged from a single weevil mummy that had been shipped from Europe. Those progeny, whose total gene pool probably came from one male and one female in southern France, soon grew into a population of millions, parasitizing as much as 80 percent of the weevil larvae in New Jersey and nearby states. In recent years, *T. incertus* has declined abruptly. In the northeastern United States, it has certainly declined because the summer larvae on which it feeds is now scarce. It is a victim of biological control's astonishing success!

All these insects are objects of the Parasite Detection Survey that the Niles laboratory (in its most recent incarnation it is the Biological Control Satellite Facility) conducts with cooperating state agencies and institutions. Where are the parasites? This is the first question asked by Tom Burger, Milt Homes, and their colleagues. When surveys indicate that a desired parasite is not in the area, the teams make releases precisely timed to the life cycles of both the weevil and its natural enemies. Later surveys answer the question of whether the parasite has established itself. These surveys are carried out not only to gather information, but also to collect a breeding stock of parasites for mass rearing and later release. In one recent year, the survey workers in Michigan, Kentucky, and Indiana collected more than 13 million alfalfa weevil larvae that yielded 7.5 million individuals of *B. anurus*. During the same year, they put back an equal number of parasites (more than had been released previously in 60 years) at 1,000 sites in 19 states.

"It's a far cry from the old days of sending a couple of workers into the fields with hand nets," Milt Holmes said. "We now use a vehicle with what we call a weevil trawler. Sweep nets are suspended on a rigid frame from the front of the vehicle. It cruises the field at eight to ten miles an hour. When the bottom board on the frame strikes the alfalfa, it whips the plant back into the net and knocks off all the insects. No, in general the farmers don't object to this technique. The vehicle does little damage to either the plants or the ground beneath. Alfalfa fields have good sod, and the vehicle stays in the same tracks, so there is really less damage than if we had a bunch of people walking around in there with sweep nets."

The survey teams use even more ingenuity in dealing with the mixed bag of debris brought back from the fields in the weevil trawler. The task here is to sort the gold from the dross, gold in this case being the weevil larvae presumably bearing valuable parasites.

"We empty the contents of the sweep net into a large pan and then pour the contents a bit at a time into a gin-trash machine," Holmes explained. "Its the kind of machine used in the South to sift through trash from the cotton gin for boll weevils. This gets rid of twigs, leaves, and weed seeds, and sorts out both the insect larvae and the adults. Then we spread all the insects and larger material on a sorting-screen table. We skim off the adult weevils with hand vacuum aspirators, but the larvae sort themselves out. Their main concern is to get away from the heat and light to which they're exposed on the sorting screen—on cloudy days we position heat lights over the table. The larvae squeeze themselves down through one of the three mesh sizes in the screens, brushing the debris off their bodies as they pass through to an inclined screen below. It's all very easy, very clean. We can just walk away and let the larvae do the job for us."

The survey teams take the weevils to the Niles laboratory, where some are dissected to calculate the percentage of parasitism. The staff rears parasites from the others (on alfalfa grown in the laboratory's greenhouses) for later release throughout the Midwest.

"We have solved most of the problems associated with rearing millions of parasites," Milt Holmes said. "In a sense we are in the horticulture business, too, growing alfalfa to raise the weevils as hosts for their own natural enemies. So we have fully equipped greenhouses with automatic watering systems, sodium vapor lights—alfalfa needs lots of light—adjustable shades to balance the sunlight and shade

Upsetting decades of biological control.

on hot summer days, even sensors that detect excessive heat or cold in the greenhouses and labs and automatically dial several home telephone numbers to alert the staff after hours.

"When the weevils have infested our alfalfa, we collect the stems and incubate the eggs in Petri dishes. After they hatch, the first instar larvae are so small that we have to make sure they can feed themselves. Nature has devised a system we call 'pressure contact,' where the tiny larvae get into the tips of the plants so they are held in tight by the new leaves—their bodies are so small compared to their mandibles that they can't bite easily unless something holds them and gives them leverage to chew. We put the larvae in jars and push leaves around them against the bottom. After they reach the second instar they don't need pressure contact and they go into rearing containers."

Taking care to synchronize the parasites with the appropriate growth stages of the weevil, the staff brings the natural enemies together with their hosts in spacious cages. In another precisely timed step, the staff packs the parasitized hosts into containers with small amounts of alfalfa and ships them off to be released in fields throughout the Midwest. It must be with some satisfaction that these devotees of biological control also contribute weevils to the Beneficial Insects Research Laboratory in Delaware, where the weevil is no longer abundant.

"In the East, the parasites have done better and better, instead of worse and worse as some of the skeptics predicted," Bill Day has said. "We estimate that biological control is saving the farmers of this region at least 8 million dollars every year—which is twice the total of what the parasite introduction program has cost since it began! But despite the fact that it's bad economics to spray at the low level of damage the pest can inflict now, there are some county agents who advise farmers to spray anyway, because they don't want to go out on a limb. They remind me of a doctor who orders all sorts of unnecessary tests and X-rays for a patient because he wants to cover himself in the unlikely event of a malpractice suit.''

Day was particularly annoyed a year or two ago when a flurry of publicity and advertising about a weevil insecticide that was said to be "safe" for parasites appeared in the farm journals.

"We tested the chemical and found that it's *death* on parasites,'' Day said. "When we inquired how the company could have made such a claim, an executive wrote back to us—he obviously didn't know anything about parasites—and included copies of their research. They had sprayed *Bathyplectes* cocoons with their product and found that it did not penetrate the hard covering. But in practice the growers don't treat the fields when the parasites are in cocoons on the ground. They treat early, when the parasites are inside the weevil and the weevil is eating alfalfa. If the chemical kills the weevil, it also kills its parasite, as well as all the parasites flying around the field.

"So the growers will use this product when the pest is at its peak and the chemical is most lethal to parasites—because if a parasite is any damn good, that's when it's going to be out there, too!''

14

MICROBIALS
Attacking the Spruce Budworm

IT IS IMPOSSIBLE TO ESCHEW SUPERLATIVES WHEN THE TALK turns to *Choristoneura fumiferana*. The larva of this insect, the spruce budworm, is North America's most widely distributed forest insect pest. The budworm outbreak that spread during the 1970s across Quebec, New Brunswick, Nova Scotia, Newfoundland, and Maine was the most serious in history. And the most intense forest spraying operation in the New World makes the budworm its target.

Unlike the gypsy moth and so many other pests in North America, the spruce budworm is a native species. It evolved with the northern coniferous forest itself, for the most part living in harmony with the forest community, occasionally undergoing a population explosion after several years of dry spring weather and a large food supply, then crashing a decade later when food grew short and parasites and diseases took their toll. In a healthy, diverse forest, it generally fits in with the myriad other organisms around it. Even budworm outbreaks are useful from nature's point of view, thinning the dense stands of susceptible fir and creating a healthier mixture of trees of various species and ages.

The budworm adult is a mottled gray moth with dark-brown markings and a three-quarter–inch wingspan. In eastern North America, it passes through a single generation annually, the females laying their 150 to 180 light green eggs in overlapping masses of a score or more eggs each on spruce or balsam fir needles in the middle of summer. The larvae hatch after two weeks. Shifting from the first to second instar without destructive feeding, they disperse and find wintering places on the trees, often in old staminate flower bracts.

The larvae emerge again in spring, usually feeding on the conifer buds. As they grow they become voracious and attack the new foliage, turning to older parts of the twig after they consume the tender needles. This is the period when budworms do most of the damage, killing many trees during successive years of defoliation. The larvae stop feeding in late June or early July and pupate within the silken nests they form by tying two or more twigs together. After 10 days the moths emerge, and the cycle goes on.

This cycle is invariable, proceeding as predictably as the earth turns. It is the other cycle, of infestation and decline, of boom and bust, that has worried foresters and the owners of the eastern pulpwood lands in recent decades, for the devastation now recurs at intervals only one-half the length of those in the era before World War II. As foresters fight what they believe to be a serious threat to the paper industry in the United States and Canada, the words "chemical treadmill" are heard as often during budworm talk as "defoliation" and "insecticides."

On a sodden morning in early June, three men stood talking on an abandoned woods road near the Canadian border in eastern Maine. They wore rain jackets and from time to time tossed small branches on a weak fire one of them had somehow started in deep ruts in the road. The fine drizzle caught and accumulated on the bare branches of the stricken firs above, to drop suddenly with little popping noises on the men's raingear.

"Will the *B.t.* stay on the trees in this weather?" one of the men, who had come to observe the annual attack on the spruce budworm, asked his companions.

The other two, both scientists with many years experience in the northern forests, looked at each other, shrugged, and then broke into nervous laughter.

"That's a question we ask ourselves every spring when it starts to rain in Maine," said John Dimond, professor of entomology at the University of Maine at Orono. "The *B.t.* is getting more sophisticated all the time, but the stickers we mix with it to keep it on the foliage are still primitive. I've seen laboratory tests, though, where 90 percent of the material mixed with stickers stayed on the foliage and only 5 percent stayed on without it. So we couldn't get along unless we used a sticker."

Dimond, in his middle 50s and just above medium height, was wearing a cap bearing the name of a manufacturer of *Bacillus thuringiensis,* the bacterial insecticide that is commonly called *B.t.* For more than a decade he has worked with *B.t.* in the laboratory and in the Maine woods, tinkering with dosages and evaluating the results in terms of both dead insects and undamaged foliage. The early tests were often disappointing. But on this day, Dimond was arguing that the most recent studies proved *B.t.* to be a viable alternative to much of the chemical spraying that was being done in Maine and eastern Canada to try to suppress the budworm.

"Just the other day I was flying over the Scientific Management Area in Baxter State Park, which is the only part of the park where the state permits spraying of any kind," Dimond said. "We've never used chemicals in there, only *B.t.* Up in the management area the forest is nice and green, but in the southern part of the park where there has been no treatment of any kind, the fir is all dead, and some of the spruce, too. It's true that we could look off in the distance and see paper company land which had been sprayed with chemicals and that was also green, but they have had to go to the expense to spray that land every year. We put on the *B.t.* in the management area and didn't have to come back with it for two years, and then only in a few troublesome spots."

Gordon Mott, the other scientist, tossed a couple of damp twigs on the fire and nodded at Dimond's account. He is a wiry man, his dark hair turning gray, who grew up across the border in New Brunswick and learned about the budworm's boom-and-bust cycle while he was with the Canadian Forestry Service. Later he became a citizen of the United States and went to work for the U.S. Forest Service. Since leaving that agency ("thanks to Ronald Reagan's early retirement program"), he has served as a consultant to forest landowners on both sides of the border.

"The great merit of *B.t.*, as far as I'm concerned, is not

just that its biological," Mott said. "After all, if you put rotenone or pyrethrum, which are both insecticides produced
from natural substances, into the environment in very high
quantities, you could very easily get into all sorts of trouble.
B.t.'s merit is its narrow spectrum. It affects *only* the caterpillars of lepidopterous insects. It does not harm the parasites and predators, or other organisms."

Like any other alternative to chemical insecticides, B.t.
has had to overcome indifference, disbelief, and even hostility from landowners and government resource managers. Although Maine and most of the Canadian provinces began to
turn to B.t. increasingly in recent years, it rankles Gordon
Mott that the official position in his home province of New
Brunswick remains cool; many government administrators
there argue that it doesn't measure up in costs and effectiveness to the chemical insecticides.

"I guess I've taken on the role of a conservative activ-
ist," Mott said with a grin. "I'm not mortgaged in any way so
I don't mind standing up and getting shot at. Most of the
members of the technical establishment involved in New
Brunswick are somewhat compelled to be a lot more conservative than that, and to let wisdom emerge slowly while contributing to it whenever they can. The fact is, in New
Brunswick B.t. has been shown to work *eminently* well.
Why, just recently they had an experiment where they used
B.t. against budworms that were in densities *higher* than
conventional wisdom said was feasible—one wonders
whether that was done *deliberately*—and it worked. It
worked surprisingly well."

Dimond nodded. "In Maine the paper companies—and
the state, too, for that matter—have come under a lot of criti-
cism from the public for using chemicals in such large
amounts," he said. "The paper companies, which are the big
landowners here, are now willing to listen to any plan that

promises protection for the forest and also has public support. They're as tired of all this spraying as anybody else—more so, in fact, because they're paying for it."

As the two scientists talked to their visitor about the budworm, they were able to illustrate their points by referring to the dripping forest around them: the tall dead firs, tufts of old-man's-beard lichen clinging to their otherwise bare branches, bleak against the bleak sky; young conifers springing up in what once was the older trees' shade, the fresh green fir a foot or two higher than the slower-growing spruce; and the occasional tiny budworm larvae beginning to feed among the new growth at the tips of the young trees, all nourishment having gone from their preferred habitat high in the mature trees.

"The landowner here was willing to let the mature fir die and then harvest it," Mott said. "But he began to get scared when the spruce started to die, too, and that's when he began to spray."

And as Mott and Dimond talked, there began to emerge a picture of this vast forest primeval (which is predominantly red spruce and balsam fir, rather than the murmuring pines and hemlocks of Longfellow's poem, though they are here, too). There are no villains in an undisturbed forest, and the modest cycle of boom-and-bust went on for thousands of years before the arrival of the European settlers. Maine and adjacent Canada had little to offer the settlers besides timber and furs, and the great white pines that grew mostly in southern Maine and along its coastline and rivers soon became relatively scarce. The lumbermen turned to the spruce forest on the state's inland plateau and sent the logs down the Penobscot, Kennebec, and Androscoggin rivers to the many sawmills that sprang up on their banks.

Because of the increasing demand for paper in the second half of the nineteenth century, manufacturers cast about for new raw materials to supplement flax, hemp, and other fibers and settled on wood pulp. (Europeans, according to one of those pretty stories that give color if not always literal truth to history, first hit on the idea after watching paper wasps chew and macerate wood to build their nests.) After some experimentation, spruce finally was settled upon in Maine as the best raw material for pulp because, in the words of an early authority in the field, it "excels in length and strength of fiber and is most readily reduced to the macerated condition."

As spruce was cut from the forest, balsam fir gained a firmer foothold and became the dominant tree in many areas. This species is more susceptible to the budworm than is the tree the insect was named for, and thus provided an almost unlimited source of preferred food for the insect throughout Maine and eastern Canada. Released from the traditional checks on its expansion, the budworm flourished. In 1912, there was a budworm outbreak apparently unprecedented in its destructiveness. Fed by forests of almost pure fir and spruce, the budworm quickly overwhelmed its parasites and predators. The outbreak persisted until 1920. According to the U.S. Forest Service, the budworm destroyed more than 27 million cords of wood, enough to supply Maine's pulp mills at the present rate of consumption for almost 10 years. Old-timers swear that a timber cruiser could walk for miles on the fallen trunks without ever touching the forest floor.

"The pulpwood industry was able to ride out the destruction until the younger trees came to maturity," John Dimond said. "They harvested the dead trees and got by on what was left of the spruce because, with the Depression coming on, there wasn't a great demand for the wood."

Disease and starvation finally caught up with the budworm, but the way had been prepared for another outbreak. The adaptable fir rapidly took over the devastated areas, creating a new forest of even-aged stands that would prove hospitable to budworms in later years.

"Canadian entomologists have figured out that there was a 40-year cycle, keyed to the time the young fir that survived an earlier outbreak grew to a size favorable for infestation," Mott said. "The cycle seems to be imposed by either of two conditions. One may be the predator-prey relationship—the prey getting scare so that the natural enemies, with nothing to feed on, get scarce, too. Then gradually the forest produces an abundance of food for the prey, which has a population explosion, but then there is a time lag before the natural enemies can catch up with it, so the outbreak goes on uncontrolled. The second scenario may make climate the controlling factor, so that the budworm rises and falls with the climatic cycle."

In 1952, just 40 years after the great outbreak, the Maritime Provinces were awash in spruce budworms. The outbreak followed its accustomed course, springing up in Western Quebec and rushing downwind toward the coast. This time the landowners had the effective insecticide DDT at their disposal. Confidently, they sprayed 200,000 acres of New Brunswick forests in 1952 and killed 99 percent of the budworms. To their consternation, however, the budworms spread over four times as much forest by the next summer. New Brunswick had stepped on the chemical treadmill.

"They were using DDT at one pound an acre and it really worked," Mott remembered. "The trees stayed green. It was like bailing out the ocean, though, and the budworms kept coming right back in."

"Maine was an exception then," Dimond said. "Perhaps because of some quirk in the wind currents, the moths did not fly into Maine but seemed to pass just north of us.

Every once in a while they would appear in some isolated pocket, but we would go in aggressively with DDT and stamp them out."

Although the outbreak subsided without the destruction left in the wake of its predecessor, trouble spots remained; New Brunswick was forced to keep on spraying, turning to shorter-lived chemicals after the government banned DDT. The interval between outbreaks, however, was cut in half. The insects returned in great force just 20 years later, fueled by the food on the trees protected by DDT. This time Maine did not escape. By the early 1970s, the state was applying chemicals to about half a million acres each year. The figure jumped to more than 2 million acres in 1975 and 3.5 million acres in 1976, the latter at a cost of more than 8 million dollars. (Even this figure was small compared with Canadian operations, with both New Brunswick and Quebec treating about three times the acreage sprayed in Maine.) Still, the outbreak was a true "tree-killer."

There was intense dissatisfaction all around. Angry groups sprang up on both sides of the border, decrying the high costs of spraying, environmental damage, and the uncertainty about what it all meant for the health of human beings. The U.S. Forest Service stopped contributing money to the program. The state of Maine also withdrew funding, and even wanted to withdraw from operating the spray program. But the landowners and the paper industry did not want to bear the legal responsibility for whatever safety problems might arise. Clearly, alternatives were needed, and that is when the state and the landowners began to look more closely at *B. thuringiensis*.

Bacteria are miscroscopic, one-celled organisms that are chiefly beneficial to humans and their enterprises. They produce enzymes that build up or break down organic compounds, and thus can be used in fermentating processes and for composting and decomposing organic wastes. Some bacteria are pathogenic, causing diseases in many species, including a variety of insects and mammals, as well as man. (Tuberculosis and typhoid fever are among those caused by bacteria.) *Bacillus thuringiensis* was first isolated early in this century from dying silkworms in Japan. In 1911, it was isolated again by a German entomologist studying Mediterranean flour moths in Thuringia, a region of Germany, and thus received its name. This bacterium has been found to cause disease in a number of insects since then, one variety being known to attack mosquitoes and blackflies.

John Dimond began his experiments with *B.t.* in the early 1970s, working closely with several Canadian entomologists. At the time, *B.t.*'s status as a weapon against the budworm was mainly theoretical. It was not present naturally in the environment in sufficient quantities to trigger epizootics.

"As *B.t.* cells mature, they produce spores," Dimond explained. "This is a resting stage of the organism, when they are resistant to unfavorable conditions in the environment. *B.t.* also produces a crystal along with the spore, and both the spore and the crystal are toxic to many caterpillars, including the spruce budworm. The caterpillars, which are the immature stages of moths and butterflies, have more alkaline digestive tracts than other insects. The alkaline digestive juices in the insect's gut work on the spores and crystals, releasing toxins that damage the gut lining. Many of the caterpillars die. Even those that survive suffer paralysis of the gut for a time and stop feeding. They grow weak and later emerge as undersized moths that die or lay fewer eggs than larger moths do. This is why a simple count of dead caterpillars doesn't tell the full story of *B.t.*'s effect.

One application may not produce a dramatic kill, but it may protect as much foliage as a higher kill by chemicals."

A number of firms now produce *B.t.* for use against a wide variety of caterpillars that attack garden crops (such as cabbage loopers, vegetable leaf miners, and tomato fruitworm). It is produced in deep fermenters, like those used to make many antibiotics, where the bacteria are provided carbon, nitrogen, mineral salts, and amino acids. When the spores germinate with the crystals, they are separated from the "broth" by centrifugal force. Low-temperature evaporation removes the excess water. What is left is a spore-and-crystal slurry that is spray-dried to make a wettable powder or, perhaps, used in other foundations, such as a liquid concentrate or a granule bait.

"The trick is to get the material onto the trees, and then into the insects," Dimond said. "It has to be thick and heavy to settle properly when it's sprayed on the trees from the air. One of the commerical formulations we use comes in what I call a dilute molasses consistency, to which the manufacturer adds a green dye so that the product always comes out a uniform color. Another manufacturer gives its *B.t.* the consistency of a kind of watered-down mayonnaise, with a buffy color. The exact ingredients that carry the spores and crystals are trade secrets."

As the drizzle turned to a steady rain, the three men retreated to John Dimond's pickup truck. They left the woods road and cruised along the highway that ran through strands of spruce and fir.

"There's no sign of budworm damage yet," Dimond said. "All this wet weather has retarded the new growth on the twigs. The larvae are in among the needles, all right, but they're still very small and not very active. In fact, it wouldn't do much good to put on *B.t.* in this kind of weather, even if the rain didn't wash a lot of it away. Budworm larvae don't feed much during periods of rain, or cool, cloudy weather. *B.t.* isn't like a chemical insecticide that just has to come in contact with an insect to kill it. The insect has to eat foliage with the *B.t.* on it for it to have any effect."

Dimond slowed the truck and pointed to two young moose standing in the road ahead. The animals, long-legged and frisky, showing their great lumpy muzzles in profile, suddenly bolted into the trees.

"Yearlings," noted Mott. "I'm not one of those who say we should just let nature take its course during these out-breaks. The forest is a valuable resource in this part of the world, with a lot of people dependent on it, and we have to protect it with the best tools we have. But it's nice to know that there's life in the woods. I know that the effect of chemi-cals is difficult to measure in scientific terms, but I was thinking of that yesterday when I was out in the woods by myself and the sun was shining. Every dragonfly was out there working the mosquitoes just as hard as it could, the birds were all singing, everything was alive. Then you go in there when a chemical has just been put on and the silence gets to you. You *know* there has to be an effect."

"Yes," Dimond nodded and laughed. "Why, there aren't even any blackflies."

The two scientists commented on the stands of trees they passed, on the severity of the damage in a specific plot, and on the impossibility at times of distinguishing between red and black spruces. ("I use it as a measure of my objectiv-ity if I'm still getting confused about what I see," Mott said.) And both men laughed again.

"I know you're thinking about what that woman said in East Machias," Dimond prompted.

"That's right," Mott said. "We were doing a field study

on budworm pheremones, putting out little squares of plastic impregnated with sex pheremones to see if they had any effect on confusing the males and interrupting their mating. A lot of work still has to be done to perfect that technique. But, anyway, we stopped at this restaurant in East Machias and told the waitress what we had been doing, and she just shook her head and said, 'Males have been confused for centuries, and that has never stopped the population from growing.''

In Topsfield, on U.S. Route 1, Dimond pulled the truck into a parking lot in front of a white frame building. "I think this building has been used as the town office," he said, "but this year the Passamaquoddy Indians took it over for the spray program on their land. We call it the Counting Mill."

Sitting around tables in a large room were 18 to 20 women, boxes of conifer twigs and stacks of charts in front of each of them. They were busily sorting through the twigs, prying apart the needles, and jotting numbers on their charts. Dimond explained to his visitor that the research crews had collected the twigs at random in areas already sprayed and brought them here to be analyzed. The women were compiling information on the number of budworms on each twig, breaking down the figures into insects dead and alive, and using a standard formula to indicate the severity of damage done to the foliage. In a back room, Dimond found several dozen glass vials holding dead budworms collected from a plot sprayed with *B.t.*; he would analyze the insects later to see if death resulted in all cases from *B.t.* or if another element was also involved.

"The plot was sprayed four days ago," Dimond said. "Apparently a lot of the insects are still alive, but they're not feeding and it looks like most of them are going to die soon, too. It's looking pretty good."

From Topsfield he drove south to Princeton, the site of

a Passamaquoddy Indian township. It is also the site of a small airport used by the spray planes. A dozen or more planes, mostly Thrush Commanders, were parked off the runway.

"A few years ago they used a variety of planes from the World War II era, PV-2s and TBMs and the like," Dimond said, "but they were pretty beat up and unreliable. The planes now are newer and mostly smaller as well, because we're spraying less land in the state and the individual plots are smaller. We're spraying fewer than 1 million acres in Maine now, and 15 percent of that is with *B.t.* Some of these planes carry chemicals, and those over there are the ones Gordon is working with to treat the Passamaquoddy lands with *B.t.* The plane that treats my plots is off spraying apple orchards with chemicals in Central Maine today, but it will be back here when the weather clears up."

Drums of *B. thuringiensis*, each marked with the trade name of its producer, stood at the edge of the field.

"We're working with a highly concentrated product now," Dimond said. "We use it neat, with no mixing, and pump it right from the drum into the aircraft. From there it's applied to the trees through the system of nozzles on the plane. The problem in the early days was that the formulations were much too dilute. They are greatly strengthened now, and since we can apply it neat, there is far less cost for handling and mixing and application. Right now we're applying the material at only 24 fluid ounces an acre, which means that *B.t.* is competitive with carbaryl, or Sevin, which has been the workhorse chemical insecticide in Maine for many years. We've got the cost down to under six dollars an acre, and now we're shooting for an even lower cost to be competitive with Matacil, which is the cheapest chemical being used against the budworm."

Dimond led the way into one of the trailers used as offices for the spraying operation in that part of the state. They were joined there by Bruce Francis, a stocky young man wearing an olive-green uniform jacket over a casual shirt and slacks. Francis is chief of the Passamaquoddy Indian Forestry Department. He supervises operations on the more than 80,000 acres of forested lands the Passamaquoddy Nation has bought so far under the terms of the Indian Land Claims Settlement Act.

Like most Passamaquoddy leaders, Francis has a strong commitment to biological control, based on his estimate of the environmental damage caused by chemicals.

"After I served in Vietnam I came home and went to the School of Forestry at the University of Maine," he said. "I was a research technician for the U.S. Forest Service for five and a half years. When the tribal council knew they were going to acquire this land they asked me to come back here and be in on the ground floor of the forestry program. The budworm is attacking our forest just like everybody else's, but all of us here are dead set against the use of chemicals. I had worked with Gordon Mott on a project at Baxter State Park when we were both with the Forest Service and I knew about his expertise with B.t. So when Gordon retired to become a consultant, I asked him to join our staff as senior forester—a good title to go with his gray hair! This year we are spraying 40,000 acres, all of it with B.t."

Gordon Mott, like Dimond and Francis, hopes that before long almost all of Maine's budworm program will be based on the use of *B.t.*

"The real challenge in managing a native population of pests—within its natural complex—with these biological materials is to get an understanding of timing," Mott said. "Is there a regular population cycle? If so, when do we intervene? I speculate that the best chance of success with *B.t.* is when the budworm population is very old—when the whole complex of its natural enemies is well established and is beginning to respond to the budworm buildup. At that point the budworm population is such that there is a good ratio of natural enemies to pests so that you can drive the pest population down. The lesson from the past is, I think, let the population go during the peak of the outbreak because you can't save much foliage by treating them anyway. If you don't spray them, you allow the natural enemies to catch up with the pests, and then you hit it when it's on the way down. That way, you still have a chance to save some trees. By using *B.t.* instead of chemicals, will you be able to work more effectively within the whole complex of natural enemies?"

Until now, Dimond and Mott have based a positive reply to that question mainly on anecdotal evidence. The latest statistics from the field, where up to 92 percent of the foliage in treated areas remained undamaged, certify *B.t.*'s status as an effective budworm killer on a wide scale. The next step is to measure the long-term vigor of *B.t.*-protected forests against those where chemicals form the first line of defense.

15

PLANT PESTS
The Water Hyacinth and the South

CULTIVATED PLANTS ARE SIMPLY WEEDS WITH THE WILDNESS bred out of them. Hybridized and otherwise coaxed by humans into sweet taste and tenderness, like circus poodles into skirts and hairbows, they offer themselves up at the table with almost audible pleas of "eat me." (Modern plant science is, in part, an attempt to breed spunk back into our fruits and vegetables.) But wild plants are versed in the art of self-defense. They evolved in the give-and-take of the natural world, and if scientists are just beginning to perceive the subtle ways in which they fit themselves for survival, farmers have known the wiles and hardiness of their plant enemies all along; they speak of them contemptuously as "weeds," just as they call the cunning wolves or coyotes who prey on their stock "vermin."

Take the lowly wild crucifers, or mustards, for instance. Like many of our most common weeds, they have adopted a variety of defenses to ward off diseases, insects, and plant-eating mammals. They contain chemicals that are unpalatable and even toxic to other organisms. Mustard oil glucosides, which have a faintly unpleasant odor as well, are the most prominent of these chemicals. But scientists also note what they describe as a second line of defense—certain alkaloids the plant evolved in response to the detoxifying mechanisms that insects developed against mustard oils. The wild mustards, as a further defense, tend to grow in association with other plants of differing chemical and physical characteristics. The cabbages, cauliflowers, and other cultivated plants derived from the wild crucifers, however, lack both the potent defensive chemicals and the defensive trait scientists call "low apparency."

"Most of our vegetable crops are derived from species which are unapparent in natural habitats," Paul Feeny, an entomologist at Cornell University, told a meeting of the American Association for the Advancement of Science several years ago. "But the effectiveness of natural plant defenses is reduced considerably by contemporary agricultural methods. When planted in monocultures, crop plants are considerably more obvious to their insect enemies, yet they lack chemical defenses appropriate for survival as apparent plants. Plant apparency could be reduced by such traditional techniques as varying crop location and field size, and by interplanting varieties of the crop that differ chemically. Levels of natural defensive compounds could be maintained or restored by plant breeding."

Some forest trees contain resins and other chemicals that are obnoxious or toxic to insects. Certain members of the bean family are covered with a mass of sharp, hooked, hair-like structures that entangle and even puncture plant-feeding insects. Several deciduous trees intersperse palatable or nutritious leaves with unpalatable ones, forcing insects, in one scientist's words, to "bustle about searching for high-quality leaves," and consequently, exposing themselves more often to predators.

Yet the natural world is constantly in flux, and adaptable insects in some cases have learned to cope with plant defenses. The larva of a sawfly that feeds on pines even processes the defensive chemicals of the tree to make itself unpalatable to birds and other predators. And everywhere throughout the natural world, individual insect species have learned to specialize in, and destroy, certain wild plants that in the insect's absence may dominate entire plant communities.

The water hyacinth (*Eichhornia crassipes*) is an unlikely villain. It has large blue-violet flowers and pretty foliage and is a relative of the common pickerelweed (*Pontederia*) that graces so many rivers and streams in the eastern United States and Canada. In much of its habitat in South America, the water hyacinth is a wild ornament, too, only occasionally becoming a pest in Argentina, Paraguay, or southern Brazil.

It is difficult to pinpoint the place and time of a pest invasion, and statements made with certainty in one decade may be revised or fuzzied up a bit in another. There is a pleasing story tied to the introduction of the water hyacinth to the United States, and it may as well be accepted as any other: During the 1890s, there was an exposition in New Orleans where some of the flora native to South America was shown, and the local people were so taken with the beautiful floating plant they gladly accepted samples of it and took them home to liberate them on nearby waterways. And so a pest was born.

"Eventually, the water hyacinth spread all around the world—including India and Australia, as well as the United States—and most times it was spread on purpose because people thought it would be such a nice plant to have around," said Lloyd Andres of the USDA's Biological Control of Weeds Laboratory in Albany, California. "But like so many other alien species, such as the mongoose, the English sparrow, or the carp, that have been introduced to new parts of the world without much planning, it soon began to increase and slid out from under our control. Now this plant clogs waterways in many places, interfering with boating and fishing and fouling pump systems. It had spread throughout much of the southeastern United States and was later carried somehow to California. Not so much in southern California, where many of the streams don't run year-round, but here in the San Francisco area it has exploded into a serious pest."

People who watched water hyacinth form solid mats over their southeastern waterways began to treat it as they would a swarm of locusts. They attacked the plants with sodium arsenite and kerosene burners at first, but the environmental upsets caused by their use were worse than the pest. After World War II, they turned to a new herbicide, 2,4-D. They met with some success using the chemical, but costs (and occasionally protests by environmentalists) discouraged the use of herbicides in many areas. It was only in the late 1960s that the struggle against the water hyacinth was handed over to biological control.

The idea of fighting weeds with natural enemies has been in the air for more than a century, although the science

demanded a high level of sophistication because of the obvious risks; insects or diseases that might be used to control
weeds are always suspect as potential crop pests. Apparently
the first example of classical biological control to be used
against an unwanted plant occurred in Ceylon (now Sri
Lanka) in 1856, when cochineal insects were brought from
India to attack prickly-pear cactus. (One of the great successes in biological control was achieved later by introducing
insects from South America to attack a similar cactus during
the 1920s in Australia, where it had become a horrendous
pest.) The United States, where there were justifiable fears
about introducing plant-eating insects of any sort, was slow
to use this technique. The first American to be active in this
field was Albert Koebele, who had moved to Hawaii after his
triumph with the vedalia beetle. Lantana, an ornamental
plant, had escaped from cultivation to become a pest on
rangeland in what was then the Territory of Hawaii. The ter-
ritorial government assigned Koebele to hunt for insects that
fed on the plant in Central and South America. Although he
found and shipped back from Mexico two butterflies, a bug,
and a lepidopterous leaf miner, these insects failed to give
the needed relief; it wasn't until additional introductions
were made in the 1950s and 1960s that headway was made
in the control of lantana.

Another famous name in the history of biological control is Harry S. Smith's, who became the chief of biological
control work in California in 1913 (and in fact, gave this
branch of entomology its name). Smith had been interested
in the use of insects for weed control since the early 1920s,
but the opportunity to put his ideas into practice did not
occur until World War II. By then the Klamath weed (*Hy-
pericum perforatum*) was thought to be an insurmountable
problem in northern California and parts of nearby states
where the failure of conventional control methods threat-

ened to put ranchers out of business. This European weed crowded out desirable forage plants and, like many successful wild plants, was mildly toxic to the domestic animals that grazed on it.

"Meanwhile, Australia had been fighting the same problem with Hypericum-eating insects introduced from England and Europe beginning in 1929," Paul DeBach writes in *Biological Control by Natural Enemies*. "Professor Smith, in California, followed the progress there with much interest through correspondence with A. J. Nicholson, chief entomologist for the Commonwealth Scientific and Industrial Research Organization (CSIRO). Smith finally obtained authorization in 1944 to import three species of beetles that showed promise against the Klamath weed in Australia. It was impossible then to consider importation from Europe because of World War II, but rather easy to bring material from Australia through the cooperation of the United States Army Transport Command. The Australian CSIRO, through Dr. Nicholson, offered to collect and prepare the material for shipment. Importation started in October, 1944, but problems were immediately encountered in changing the timing of the life cycle so the beetles would be in phase with the Northern Hemisphere seasons. Two species of *Chrysolina* that were in the summer resting state of aestivation responded rapidly in California to fine mist sprays of water to become active and lay eggs within three weeks. The third species, an *Agrilus,* was lost. After starvation tests in quarantine on a variety of economic plants, the beetles were ready to be released in the field."

The two beetles imported by Smith had passed the test of a plant-feeding insect recruited for biological control: They fed only on the target species or closely related plants of no economic or esthetic value. (A minor fuss developed in California years later when a closely related ornamental *Hy-*

pericum began to be widely planted by landscape gardeners in California, in complete disregard of the beetles' presence there, and naturally came under occasional attack.) The imported beetles devastated the Klamath weed, the fall and winter defoliation of its rosettes affecting plant root development so that they could no longer compete with other plants. The weed died back, permitting the return of the bunchgrasses, clovers, and other desirable forage plants. In gratitude, the region's stockmen raised a monument to the beetles.

Since then, the use of insects to fight weeds has lost its novelty in the United States, especially in the western states. They have been turned loose against various thistles, tumbleweeds, and puncturevine. The USDA has established laboratories to collect plant-feeding insects in Italy and Argentina. Lloyd Andres, although he has always worked to control western weeds, now finds himself in the USDA's California laboratory monitoring what used to be considered a pest of the southeastern states, the water hyacinth.

"All weeds have natural enemies somewhere in the world," Andres said in his office at Albany. He is a tall, soft-spoken man with a mustache and thinning, light-colored hair. "What we're trying to do is to move these natural enemies from one area to another to complement the climate or other environmental stresses that already act on the weed, even in areas where it is a pest. In other words, we put that little extra bit of pressure on the weed that may spell the difference between it being a pest or just another plant in the community of plants."

Andres got his start in the more usual area of insect control. "I worked as a technician with DeBach at Riverside, raising parasites on the citrus red scale," he said. "I came to Berkeley to get my degree, and then was back at Riverside when I learned of an opening in weed control with the USDA. I had been interested in this subject ever since I heard a talk by the man, curiously enough, I was to succeed here, James K. Holloway. In speaking about how perfectly insects are synchronized to their host plant, he described a weevil that attacks gorse, a spring shrub that forms impenetrable thickets in some areas of California and Oregon. This weevil, which has only one generation a year, destroys the plant's seeds. The overwintering adult weevil stays dormant until just before the plant buds, when it emerges and feeds on the flower petals. This feeding nourishes the female's eggs, which she then lays inside the forming pods. The hatching larvae consume the seeds. Later, after the larvae have matured, pupated, and transformed to the adult stage, the pods open, releasing the new adults, which feed and then prepare for overwintering. I found the whole story fascinating. Well, I applied for this position, and spent a number of years chasing down insects in Europe before returning to California."

Meanwhile, the USDA had accepted the task of trying to clear waterways clogged by aquatic weeds, using the time-honored techniques of biological control as their major weapon. Its entomologists first took on alligatorweed (*Alternanthera phylloxeroides*) at the request of the U.S. Army Corps of Engineers. This plant had entered Florida and Alabama in the 1890s, about the same time the water hyacinth entered. It is a perennial that reproduces only by vegetative means in North America, according to Andres, and a new infestation can sometimes arise from a single stem node. Alligatorweed spread up the Atlantic Coast to North Carolina and along the Gulf Coast to Texas, clogging waterways and ruining water quality. The Army Corps of Engineers, which is responsible for keeping navigable waterways open, had little success in controlling it with herbicides. It is an emergent plant (contrasting with water hyacinth, a floating plant), rooted on the bottom and reaching the surface on long hollow stems to form its huge mats. Herbicides withered the surface foliage, but the plants sprang up again almost immediately from their roots in the bottom mud. In 1960, the Corps called for help.

"USDA sent an entomologist to search for natural enemies in South America, where alligatorweed was thought to be native," Andres said. "Finally, after several years and a lot of intensive tests, the agency released 266 adult flea beetles, identified as *Agasicles,* near Jacksonville, Florida, in 1965. These insects feed and reproduce on the leaves and stems of alligatorweed. By the next year, the observers who went to the area estimated there were hundreds of thousands of *Agasicles,* and pretty soon the site was essentially cleaned out. It is believed the insects were so successful because they destroyed the new spring growth at a time when the weed mat's carbohydrate reserves were at a low level. The insect spread to other areas, and the agency suddenly had a successful biological control project."

The Army Corps of Engineers was still making do with herbicides in its struggle to contain water hyacinth, but it was losing. For a while, it was inclined to let native organisms, from snails to manatees, put pressure on the dense mats of water hyacinth. (Conservationists were eager to let manatees, those rotund, endangered aquatic mammals old-time sailors are said to have confused with mermaids, prove their worth to society by attacking an economic problem; but the animals found other plants in their habitat equally digestible and the conservationists finally admitted that the manatees would have to be loved for themselves alone.)

Called in for help once more, the USDA turned to its South American laboratory, which happened to be located in the hyacinth's range in Argentina. Collectors found a weevil (which later turned out to be two different species) feeding energetically on water hyacinth near Buenos Aires. These insects (genus *Neochetina*) are well adapted for spending most of their lives in the water, being covered with a layer of dense water-repellent scales. They lay their eggs in the host plant's petioles (leaf stems). The larvae burrow into the petioles at first, feeding on the inner tissues and often causing the leaves to wilt or even fall into the water. As the larvae grow larger, they turn toward the center of the plant, where they excavate small pockets. Later, the larvae (which are so host specific that they will not develop on other plants) leave the stem and pupate underwater in cocoons attached to the plant's upper root zone. When adults emerge, they damage the leaves by scraping off the epidermis with the mandibles at the tips of their long snouts.

The introduction of the weevils into the United States was greeted enthusiastically by almost everyone concerned with water resources in the Southeast. Within four years of their arrival in Louisiana, that state's area of water hyacinth shrank from over 1 million acres to 350,00 acres, or less. Yet here and there water managers in the South, addicted to the instant gratification to be gained from the use of chemicals, turned once more to herbicides.

"When extensive acreages of water hyacinth are treated with relatively fast acting herbicides, all of the immature stages of the insects are eliminated along with the plants," Ted D. Center of the USDA's Aquatic Plant Management Laboratory in Fort Lauderdale, writes in *Aquatics*. "When plant growth occurs, few insects are present to suppress it and reinvasion occurs rapidly. A few adult weevils may find these plants and begin to feed and lay eggs but, because the plants grow so quickly and the weevil population so slowly, the plants are again a problem before sufficient numbers of weevils are present to have an effect. It then becomes necessary to use the herbicide again, and again the insects are eliminated. This repetitive cycle is the primary reason that the weevils have not been effective at highly managed sites."

Herbicides, when used properly and sparingly, may work for the weevils, putting the plant under greater stress and allowing its natural enemies to catch up with its growth. Researchers at the Fort Lauderdale laboratory have also experimented with other ways of helping the weevils. Adding plant-feeding fish called white amurs to a pool of water hyacinth infested with *Neochetina* weevils, they found that the two creatures worked splendidly together. The weevils attacked the leaves and stems, and the fish ate the roots as well as the leaves, so no part of the plants was left untouched. Above all, timing is important: getting the insects in sufficient numbers on a plant when its carbohydrate reserves are low. Research completed at the USDA's Argentine laboratory shows that the *Neochetina* weevils keep water hyacinth under tight control there, though they themselves are constantly under attack by a variety of natural enemies, such as nematodes, mites, and predaceous beetles. Theoretically, then, the weevils should flourish in the United States, where they have no natural enemies, as far as is known.

Although the water hyacinth was present in isolated sloughs in California for a number of years, it did not become a problem in northern California until the late 1970s. Then it came into its own, erupting from the sloughs to block the entryways to marinas and to choke the pumps that lift water from the delta region to send it on its way toward southern California.

"It certainly became a lot more noticeable," said Lloyd Andres. "It was often growing more than three feet high, and when there was a lot of rain you could see plants being washed down San Francisco Bay, going out to sea. We couldn't use chemicals to kill the water hyacinth, since so much of the water in the delta is used for drinking or irrigation. We asked our laboratory in Florida for suggestions and they sent us the two *Neochetina* weevils and a moth."

Before an insect can be imported into the United States for use in weed control today, all the information compiled on it is studied at great length by a committee that represents various federal agencies; it is the Working Group on Biological Control of Weeds. But each state holds veto power over what may be introduced within its borders and may reject an insect if its scientists believe there is still cause for concern.

"We took one of the weevils to the University of California at Davis after the state looked over the testing data and cleared it for introduction," Andres said. "The USDA has an aquatic research laboratory there. Our people got the weevils breeding in tubs of water hyacinth, and when the weevils were just maturing we set out the infested plants in the delta near Tracy. Biological control usually takes a while to show results. But after a while I went back to the first release site where I could see a slight but very definite depression in the mat of vegetation. I pulled up some plants and found the weevils. I was pretty excited at being able to see an impact so quickly. However, it's not as humid here as it is in the Southeast, and the plants are already under a certain amount of stress. So, in California, if these insects are left undisturbed, they may very well push water hyacinth toward a swift collapse."

16

A PLAGUE OF LOCUSTS
The Search for Nature's Solutions

GRASSHOPPERS BELONG TO THE ORTHOPTERA (MEANING "straight-winged"), an order that also includes the crickets, katydids, cockroaches, praying mantids, and walkingsticks. Adult "hoppers" are large animals, as insects go, many North American species being an inch or more long, and one of them reaches two and one-half inches with a wingspan of four inches. They are physical in the sense that we apply the word to pro football linebackers, being encased in a hard-plated shell, much like a lobster's, packed with powerful muscles that enable them to leap 20 times their length and make long flights while swarming. The migratory grasshopper (*Melanoplus sanguinipes*) is known to fly 25 to 30 miles a day.

"In 1938 swarms of this species flew from central South Dakota to central Montana, a distance of about 500 miles, in three weeks," wrote USDA entomolgists J. R. Parker and R. V. Connin. "Between 1874 and 1877 great swarms of the Rocky Mountain grasshopper originated in Montana and migrated eastward to the Mississippi Valley and southward to Texas. Each swarm completed the migration in a single season."

A close look at the head of an individual hopper convinces the observer that this is a formidable beast. There is a large compound eye on each side of the head, composed of thousands of facets that receive and reassemble a single picture in the grasshopper's brain. The powerful jaws are edged with sharp teeth. However, according to Parker and Connin, up to 90 percent of all the grasshopper damage inflicted on North American agriculture is said to be the work of just five of the six hundred native species.

Grasshoppers are among the minority of insects that do not completely metamorphose in their life cycles. Young grasshoppers look much like their elders, but are smaller and lack wings. There is only one generation a year in the northern United States and Canada, the adult females laying their eggs in the summer and fall, and in most species, the eggs overwinter in the ground and hatch the following spring.

There are no larval or pupal stages in grasshoppers. The young hopper simply eats and grows, the hard outer shell splitting down the back and being cast off when the insect outgrows it. This molt occurs five or six times in the hopper's development.

"The last molting process is an event that any nature lover will find absorbingly interesting," Parker and Connin wrote. "It is a common sight on warm days in the early summer when grasshoppers are abundant and first getting their wings. The grasshopper about to acquire wings hangs head downward, gripping with all six feet whatever it is resting on. The old skin then splits down the back and the emerging adult, aided by gravity, slowly pulls its feelers and legs from their encasements. As soon as it is free, it turns and faces upward so that the crumpled damp wings will hang straight down as they unfold. Slow, pulsating movements of the body force blood and air into the wings until they are inflated. Then the wings begin to dry, stiffen, and take on their normal colors. Finally the underwings are neatly folded and tucked under the protecting border and narrower outer wings."

After mating, the females oviposit in the earth. The abdomen, which is composed of segments linked by elastic membranes, stretches to two or three times its normal length as the hopper forces her eggs into the hole. Before leaving, she closes the opening in the ground with scraping movements of her hindlegs and abdomen. The eggs, cemented into pods of fifteen or twenty, and sometimes many more, endure the winter's cold as well as their progenitors adapt to the intense heat of plains and deserts. During infestations, the eggs are present in enormous numbers. Observers discovered a 20-acre field in Utah where every square foot held an average of 25,000 eggs, or 1 billion to the acre.

To walk into a western meadow infested with grasshoppers is to experience the sensation of splashing through a shallow pool, hundreds of hoppers rising about one like bursts of spray. They are of all sizes and colors—brown, green, lemon-yellow, bright orange. Over there, in a bare patch among the grasses and wildflowers where red ants had built a mound, dark-gray and white hoppers with reddish legs and white antennae are pumping eggs into the earth through their ovipositors. An infestation—a grasshopper "plague"— seems to be a spectacle of perpetual motion, without beginning, without end.

That the grasshopper, or its alter ego the locust, has come to symbolize a destructive pest is understandable. In the almost unimaginable numbers that make up their swarms and in the devastation they sometimes leave in their wake, these insects figure prominently in historians' accounts of human misery. Their depredations circle the earth, particularly in the drier regions. The villain mainly identified with these large-scale plagues is the locust, a name sometimes loosely and confusingly used to cover a wide range of grasshoppers, and even a species of cicada (an insect in an entirely different order) is called the seventeen-year locust. Perhaps the most apt distinction was offered by Joseph Gentry, a USDA specialist on grasshoppers. "A locust," Gentry said, "is simply a grasshopper gone crazy."

More technically, entomologists generally describe a locust as a short-horned grasshopper, principally of the family Acrididae, which has short antennae. (The long-horned grasshoppers and katydids, which sometimes have antennae longer than their bodies, belong to another family.) Moreover, the term locust applies especially to those Old World species that appear in two phases, one solitary, the other gregarious.

Locusts, like wolves and vampires, were among the demons that filled the nightmares of our ancestors. Their likenesses were carved on temples in ancient Egypt. To desert people they were "the teeth of the wind." Locusts helped to bring Pharaoh to his knees in his dispute with Moses, arriving in Egypt (with divine guidance) one morning on the east wind. We learn the results in Exodus: "They covered the face of the whole earth, so that the land was darkened; and they did eat every herb of the land, and all the fruit of the trees which the hail had left; and there remained not any green thing in the trees or in the herbs of the field, through all the land of Egypt."

The most notorious of its tribe is the desert locust (*Schistocerca gregaria*), the scourge of the Bible and the Koran. Entomology overflows with astounding examples of ingenious behavior and adaptation, but few animal strategies have had such a far-reaching effect on the environment as that of the locust. Even the experts had not observed the phenomenon, and it was many years until it was detected and eventually described in 1921 by the internationally prominent entomologist Boris Uvarov. Before then, the major problem in unraveling the life histories of the desert locust and its destructive relatives was the whereabouts of those insects between plagues. They seemed to disappear from the face of the earth. Uvarov was studying what had been considered to be two species, the migratory locust (*Locusta migratoria*), periodically an awful pest, and an inoffensive insect then called *Locusta danica*. They didn't look alike, nor did they behave alike. In fact, entomologists had noted in the field that if *L. danica* happened to find itself in the path of swarming migratory locusts, it scampered out of the way instead of joining the swarm.

But eventually Uvarov noticed that under certain conditions the "two species" blended in a remarkable manner and the truth dawned on him. He had been studying two phases of a single species. He and other entomologists went on to elaborate the alternating guises under which the desert locust and other pests in this grasshopper family appeared to the world.

The desert locust, in one of its phases, hatches from the egg as a small, wingless, brown or greenish hopper that generally shuns others of its kind. It is said to be in its solitary phase. As the hopper molts and matures, it grows wings and may live a comparatively uneventful existence. But under certain weather conditions not fully understood by scientists, a dramatic change takes place in those solitary hoppers. As the population increases, the insects crowd together and begin to travel in swarms. The crowding stimulates such physiological changes as the lengthening of the wings and the enlarging of the head. If no natural or human disturbance checks the swarm's growth, it will increase perhaps 100-fold with each annual generation.

Within a few years, the swarm may include billions of voracious locusts, stretching for hundreds of miles, carried by winds that naturally converge on moist regions and provide the swarm with fresh greenery. The locust plague rages, bringing misery and starvation through great swaths of Africa, the Middle East, India, and Pakistan. Once a locust plague gains momentum, there is little that can be done to control it. The sheer numbers of locusts simply overwhelm such natural enemies as parasitic insects and birds. A stork, for example, was found to have nearly 1,500 locusts in its gut, but such predation makes no dent in the incredible hordes. Eventually, disease, starvation, and—most important of all—weather patterns break up the swarm. The locusts "disappear," to be replaced by small numbers of comparatively harmless solitary grasshoppers. The wind has lost its fearsome teeth.

Because of the highly migratory nature of the desert locust, such remedies as classicial biological control have not turned out to be practical. The only defense at this time is to

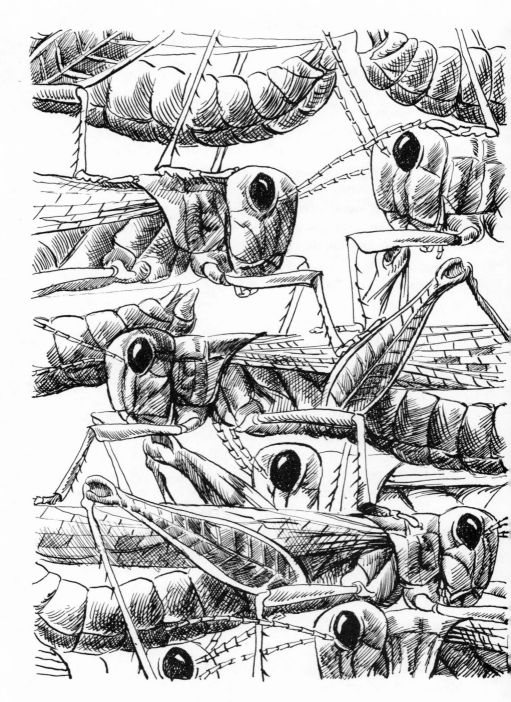

A plague of grasshoppers.

attack the swarms with chemical insecticides, which may hold down numbers but do not end genuine plagues. The grasshoppers that become pests in Western civilization have always presented farmers with a considerable problem, too, but on not quite the same scale. In his book, *Man's Plague?*, the entomologist Vincent Dethier points out that insect devastation played a much smaller role in European civilization than elsewhere, never being mentioned in Europe's agricultural chronicles, although serious damage must surely have occurred from time to time. "The truly great decimations of crops, the monumental catastrophes, have been brought about, not by insects (locusts excepted), but by weather and by plant pathogens," Dethier wrote.

Insects were a problem for European settlers in North America almost from the beginning. Eventually, the settlers created many of their own problems by accidentally importing, with other cargo, various Old World insects to America, where they flourished in the absence of their natural enemies. But much of this lay in the future. Already on hand, of course, were native insects, most species of which were benign, although the settlers were able to recount some entomological horror stories.

One report tells of grasshoppers devouring crops around Nahant, Massachusetts, until the residents "formed a line and with bushes drove the grasshoppers into the sea by the millions." Three thousand miles away, the hoppers invaded California's mission gardens. These invasions troubled the padres and rancheros for many decades, causing stern measures to be taken against the hoppers, as at the Mission of Santa Clara where, according to an early account, "Padre Jose Viadere fired the pastures, and getting all his neophytes together made such an awful noise that those which were not killed by the smoke and fires were frightened off so thoroughly as to save the grain crops and the mission fruit gardens."

Still later, Mormon historians described the dramatic incident of the "crickets" (in reality, wingless hoppers) that threatened to destroy the settlers' first harvest in Utah in 1848 when flocks of California gulls providentially arrived to devour them. But these infestations, on the whole, were local and damage was restricted to small areas. American grasshoppers did not really have an opportunity to cause widespread destruction until after the Civil War when settlers began to farm the subhumid prairies and the Great Plains near the one-hundredth meridian. This was the point at which John Wesley Powell and other government scientists warned that the lack of water would cause farmers serious difficulty.

The settlers kept coming and, for a time, they enjoyed bumper crops. In 1873 they planted their crops, looking forward confidently to the harvest, but as chess players say, "Before the ending the gods have placed the middle game." Drought struck suddenly, followed by grasshoppers that poured into parts of Minnesota, Iowa, and the Dakotas to feed on the corn, wheat, and other crops farmers had unwittingly planted for the swarms' sustenance. The hoppers returned in 1874, spreading ruin from Texas, through Kansas, Nebraska, and the Dakotas, to the Canadian border.

Gilbert Fite, in *The Farmers' Frontier: 1865–1900*, has described the plague and its aftermath. Thousands of families were destitute, deprived of both the harvest they had counted on to get them through the hard winter as well as the cash to buy the necessities of life. Ironically, it must have seemed to the stricken settlers that the only items left to eat were the hoppers themselves, for one observer wrote that when the insects were at their peak a person had to keep his mouth closed to "eschew an involuntary morsel."

After a year's respite, the hoppers struck again in 1876. All remedies proved fruitless. The governors of the afflicted states met to discuss the problem and called for federal aid, which was slow in coming. One man proposed that prayer would be the most economical approach, but nothing turned the hoppers aside. The plague finally burned itself out in 1878.

Slowly, farmers and entomologists began to raise defenses. In 1885, D. W. Coquillett, the entomologist who had assisted Albert Koebele in the campaign against the cottony-cushion scale, devised a poison bait consisting of wheat bran, sugar, water, and arsenic. For more than half a century, this bait was the standard remedy against grasshoppers. In modern times, farmers turned increasingly to chemical insecticides applied from the air. Yet occasional serious outbreaks of grasshoppers have occurred in the modern West, especially during the Dust Bowl days of the 1930s, and again in both the 1950s and the late 1970s. Understandably, the word grasshopper induces uneasiness, and sometimes panic, among western farmers and ranchers. Hoppers are able to gain a foothold on rangeland, which is comparatively undisturbed, and so build up their populations from year to year. The USDA entomologists once estimated that even during a light infestation, when hoppers averaged six or seven to the square yard, those insects consumed grass on each ten acres at about the same rate as a cow.

John E. Henry is a specialist on the hundred or so species of rangeland grasshoppers in the western and northcentral states. His headquarters is in Bozeman, Montana, where he earned his doctorate in entomology from Montana State University in 1969 and went to work with the USDA's Rangeland Insect Laboratory. When Henry began his work, the long-lasting chemical insecticides, such as toxaphene, were the standard control substances used against grasshoppers. But toxaphene and the others went the way of DDT, being banned for most uses, and some agricultural experts have never been satisfied with the control gained from their less persistent chemical replacements. In any event, none of the chemicals provided any control from one year to the next. The search was on for alternatives.

Henry's interest lay from the beginning in pathogens—organisms that cause disease in their hosts. Grasshoppers, like all numerous and wide-ranging insects, are victimized by many kinds of natural enemies—birds, rodents, larger mammals (coyotes find hoppers an abundant and nutritious food), and other insects among them. Parasitic insects, especially certain flies, attack grasshoppers. All of these enemies take their toll and undoubtedly help keep most grasshopper species from becoming pests. Yet they have little impact once the great swarms develop, and at those times more powerful forces take over and finally devastate the hordes. Weather is one of those forces, disease another.

"As with most insects, grasshoppers are hosts for a number of fungi, bacteria, protozoa, and viruses," John Henry has said. "Some of these microorganisms are pathogenic and therefore are potentially useful as agents for microbial control. However, financial and physical limitations prohibit the simultaneous development of all potential pathogens so it is necessary that one or a few be selected for major emphasis."

Henry's observations and his search of the literature turned up two kinds of viruses and five species of protozoa that seemed to be able to kill large numbers of grasshoppers. A chief advantage of viruses is that they are *virulent*. (The word *virus* in English originally referred to the venom of a poisonous snake or other animal.) Henry, however, knew that among the rangeland grasshoppers there were about 15 species in several genera capable of causing various degrees of economic damage. The viruses and a couple of protozoa he had considered as candidates attacked a very limited number of grasshopper species.

Henry and his colleagues at Bozeman finally settled on a protozoan, *Nosema locustae,* that at first had seemed to offer rather feeble credentials for grasshopper control. (The scientific name can be translated loosely as "grasshopper sickness.") There are about 30,000 species of protozoa, one-celled organisms that are considered the simplest form of animal life and most of which are invisible to the naked eye. Many kinds are free-living in water, although some are parasitic on larger animals and may harm them. (Malaria and sleeping sickness are among the diseases caused by protozoa in humans.) The genus *Nosema* is traditionally assigned to a group of protozoa known as the microsporidians. These organisms reproduce by spores, each of which contains, in turn, a single organism that eventually may, through growth and division, produce hundreds more. Certain species of *Nosema* are known to cause disease among several major insect pests, including the European corn borer and the spruce budworm.

But how effective could *Nosema locustae* be in controlling grasshoppers? The species was not especially common in nature. Moreover, E. U. Canning, the entomologist who originally described the species in England, was doubtful about its possibilities for biological control. She pointed out its comparatively low virulence in adult grasshoppers and the difficulty in mass producing spores for release into the environment.

Yet as Henry went through the list of the qualities required of a pathogen in biological control, he decided that *Nosema* measured up better than any of the other candidates. Although this protozoan attacked a wide range of grasshoppers and was able to survive in the harsh climatic extremes of the western and northcentral states, it did not harm non-target animals. Henry tested the pathogen in long-term feeding trials with rats, and in "acute" feeding trials with trout and sunfish, without finding any signs of toxicity. There was no skin irritation when it came in contact with rabbits and guinea pigs, no ill effects when inhaled in large amounts by rats. Honeybees, which are so vulnerable to chemical insecticides, proved immune to *Nosema*. Because pathogens used in insect control must meet the safety standards set for "insecticides," *Nosema* not only could be registered for use against grasshoppers, but was infinitely preferable from a safety standpoint to any registered chemical.

Having established *Nosema's* safety to vertebrates and beneficial insects, Henry and his colleagues went on to prove that this protozoan's reported handicaps were either illusory or might not seriously interfere with its function as a biological control agent. It does not kill large numbers of grasshoppers quickly on contact, as does a chemical insecticide. An insect must ingest protozoa before it becomes diseased, and even then the pathogen works slowly. Yet this drawback is balanced by the protozoan's persistence. Because the infected grasshopper lives for many days, it has an opportunity to pass on the disease to other hoppers and even spread it to distant areas. There is no need to kill a very high percentage of grasshoppers in any region, because ranchers do not consider them an economic burden at densities of under ten hoppers per square meter.

"An objective of microbial control is to provide some short-term control, enough to reduce densities to or below the economic threshold," Henry pointed out. "Another is to provide long-term control that thereafter will maintain densities at subeconomic levels and reduce the frequency of outbreaks. An ideal approach to managing the densities of grasshoppers would be to apply a microbial agent during the early stages of the density cycle, when densities are lowest, and prevent significant increase. However, such an approach may not be realistic because survival potentials among grasshoppers are highest during low density periods when loss to natural enemies also are at the lowest level. Conceivably reductions of 70 percent due to a microbial agent, or any other factor, during low densities might not be as effective as 20 percent reduction when densities are higher and there is more pressure from other controlling factors. Also, even if this were effective, implementation would be difficult because landowners and operators generally are reluctant to cooperate in grasshopper control programs prior to the existence of threatening or damaging densities. Land managers are familiar with the variability in rangeland conditions and the severe fluctuations in densities of grasshoppers, and they often prefer to withhold control action in the expectation and hope that some natural calamity will eliminate the problem."

Through field tests, John Henry and his fellow researchers learned that *Nosema* could take hold in mixed grasshopper populations. One of their projects was a five-year study in an area of approximately 200 square miles in an isolated mountain valley at 5,000 feet near Fairfield, Idaho. This region, used for grazing livestock as well as growing alfalfa and winter wheat, is bordered on two sides by mountains, on the south by foothills covered with grass and sagebrush, and on the east by sagebrush flats. There the observers were able to track the progress of the disease.

Several species of grasshoppers became infected early

in the season. The protozoa, growing slowly (it takes about 13 days for the first generation to develop in the laboratory) and restricted to the body fat, competed with the host for its energy reserve. Although the vigor, and later, the fecundity of the hoppers were reduced, they were able to survive as an intact population to infect other species coming along later in the season. An important element in the transmission of the disease was cannibalism. Dying, the insects were devoured by other hoppers and thus the disease was passed on. Later studies in the laboratory showed that the offspring of infested females were sometimes infested also, suggesting the possibility of trans-ovum transmission of the disease. By the very nature of the disease's progression, the dominant species of grasshoppers in the valley were the ones that suffered the highest overall percentages of infection.

Henry was also able to mass-produce *Nosema,* using the grasshopper (*Melanoplus vivittatus*), a robust insect that tolerates extremely high concentrations of spores. An individual, under laboratory conditions, yielded 3 billion or more spores, so that the researchers treated three or four acres of rangeland with the spores taken from a single large hopper. Because, at the present state of the art, *Nosema* seems to lose its effectiveness after being stored for any length of time, the staff at Bozeman produces the spores through the grasshopper season and thus is ready to meet any unexpected outbreak.

The process, according to Henry, is "relatively easy." Staff members rear grasshopper hatchlings 200 or 300 to a cluster in laboratory tubes, through the first molt. Then they are transferred to large screened cages, where they are fed lettuce sprayed with *Nosema* spores. After four days, the pathogen has infected 99 percent of the hoppers. The spores reproduce as the hoppers grow to adults and, if the staff did not sort them two or three to small rearing vials, the stronger hoppers would begin to cannibalize their debilitated cage mates.

"The grasshoppers are now thoroughly impregnated with the disease," Henry said, "And we're ready to collect the spores. First, the dead hoppers are crushed in a wheat mill, suspended in distilled water, and agitated to release the spores. Then the material is passed three or four times through cheesecloth and cloth screens to remove large fragments of tissue. Finally, the spores are concentrated and cleaned by centrifugation and stockpiled in distilled water."

Years of experiments have not uncovered a better way to distribute the spores than on the wheat bran flakes prepared a century ago in California by Coquillett. They are scattered in grasshopper-infested areas, either by turbines mounted on trucks or by low-flying aircraft. Henry and his colleagues have found they achieve much greater accuracy when they drop the relatively bulky bran bait from the air, rather than by water sprays, which tend to drift.

Also, we have found that full coverage is not required in the application of *Nosema* on wheat bran such as is required for suitable control with insecticides," Henry said. "Nor is re-treatment of most areas necessary. Furthermore, chemical insecticide applications are restricted to early morning hours when the temperature is such that thermals are not produced. The thermals are updrafts of warm air that cause loss of spray droplets through drying and drift. Normally this restricts aerial applications to one load each suitable day for insecticide applications in the western United States. In contrast, we have achieved excellent results in applying wheat bran with spores throughout the day in high temperatures and winds that would prohibit spray applications. This allows for increased efficiency in the use of airplanes and their crews, which lowers the cost of application.

"We've seen *Nosema,* used early in the summer, kill 50 to 60 percent of a hopper population within four to six weeks, and remain viable over the winter to infect the following year's population. With improved techniques, we may get a 75 percent kill. This is proving to be a relatively inexpensive way to keep grasshopper populations below pest levels."

Besides being effective and inexpensive, this microbial control has another advantage. It works well in conjunction with some of the less destructive chemical insecticides. Malathion, for instance, can be mixed with *Nosema* in bran baits at very low levels to provide extra potency or, when the hoppers undergo a cyclical upsurge, the insecticide can be sprayed on fields to bring down the population to where the pathogen once more functions effectively.

Nosema alone is a powerful and dependable tool for grasshopper control. As John Henry and his colleagues are proving, it is also an environmentally acceptable addition to the stock of devices required as North America turns increasingly toward the new strategy of integrated pest management.

17

THE
COTTON CONUNDRUM
The Implacable Boll Weevil

THE COTTON WEEVIL (*ANTHONOMUS GRANDIS*) LIVES IN THE odor of infamy. No other animal in North America has aroused in humans such a keen sense of personal injury, for it struck without warning at the heart of the Deep South's economy, and its very way of life, at a time when that region had not recovered from the devastation of the Civil War.

An adult boll weevil is an outlandish creature to begin with. Weevils, which belong to the order of Coleoptera, or beetles, make up one of the largest families of insects. Scientists have already described 40,000 species of weevils, and there are untold numbers out there waiting for discovery. The reason for their boundless diversity is that they tend to specialize, each kind claiming a specific plant as its own, so that many of our most common trees, shrubs, and herbaceous plants are host to a distinctive weevil. These insects are often known by the names of the plants whose nuts, fruits, or other vital parts they infest: acorn weevil, pepper weevil, strawberry root weevil, or rice weevil.

Structurally, the adult cotton boll weevil is somewhat typical of the family. A grayish-brown snout beetle, from one-eighth to one-half of an inch in length, its rounded body appears rather fuzzy because of its covering of short, fine hairs. Like most other weevils, its most distinctive feature is the long, curved snout, or rostrum, which is one-half the length of its body. The snout bears the antennae, which may be folded back in grooves along its sides; at the tip of the snout is a set of powerful jaws.

Boll weevils spend the winter in diapause. In early spring, they emerge from the leaf litter or other debris in which they have hidden at the edges of fields and seek out young cotton plants. The feeding by these adults is not especially harmful to the crop. They consume some of the tender growth at the tips of the plants, but their main food is the protein-rich pollen developing in the "squares," or small buds, that will produce the flowers and bolls.

The true deviltry of this emerging generation lies in the provisions it makes for its offspring. After less than a week of feeding and mating, the female weevil begins to look for places in which to lay her eggs. Perhaps this species should really be called the cotton square weevil because the female generally oviposits in the tender young flower buds, rather than the tough bolls that are the plant's seed capsules.

Having selected a square or a newly formed boll, the weevil punctures it with her jaws. Turning, she inserts her ovipositor through the opening and lays an egg inside the square. After she withdraws her ovipositor, she seals the opening with a sticky secretion.

Three to five days later, the egg hatches into a pale legless grub that bears little resemblance to its parent, aside from its powerful jaws. This is where the plants (and the cotton farmer) come to grief. The square turns yellow, its bracts flare, and soon it drops from the plant. The larva feeds for a week, and perhaps for as long as two weeks, on the tissues of the square that envelops it. The adult female may have laid 200 eggs, usually one to a square or small boll, and thus aborted flowering and fruiting on a number of plants.

Unchecked, the weevil population explodes in a cotton field. After four to six days as a pupa, the insect develops into an adult, chews its way out of the square or boll, and begins feeding on other fruiting forms. It mates within a few days and the cycle begins again. An entire life cycle is completed in about 18 days under optimum conditions during the growing season, and six or seven generations may be produced annually.

As the plants mature and cold weather comes on, the final generations of the year prepare for diapause. The weevils grow fat, feeding on the squares and small, soft bolls remaining at the tops of the plants. Their reproductive organs atrophy and the water content of their bodies dwindles. They abandon the cultivated field for its brushy edges, where they find shelter and oblivion until another spring.

J. Knox Walker is a tall, gray-haired man, an enthusiastic hunter of ducks and geese, and Professor of Entomology at Texas A&M University. He sat in his office among textbooks and research reports, looking back on nearly a half-century of intimate acquaintance with the cotton boll weevil.

A cotton boll worm at work.

"I was born right here on the campus," he said. "My father was a civil engineer, and in those days there were only about 300 people living at College Station. The Brazos River system is just to the west of us. The people over there grew a good grade of cotton—they were kind of the upper crust of local society—and when I was a kid I used to go there and watch the entomology experiments. They were using calcium arsenate against the boll weevil in Texas then, and it was very exciting to watch those old planes come in low and dust the fields."

The proximity of the campus (which has always been the center of the state's entomological research) to the Brazos Valley affected the way Texans confronted the problem of the boll weevil. Results of the work performed there, both on estimates of the threat posed by the insect and the methods of alleviating the damage, were extrapolated to the rest of Texas.

"Do you know anything about climbing Mount Everest?" Walker asked his visitor. "George Mallory made one of the early attempts on the mountain, and lost his life doing it. I remember reading that when his party originally got permission to go through Tibet to reach Everest, they couldn't even get to it because there were all those lesser mountains around it. So Mallory wrote down in his diary something like this: 'It will be necessary in the first place to find the mountain.' Anyway, in entomological terms it was really necessary in the first place to run the kind of experiments to find out what you could do about the boll weevil in the various regions of the country. We had learned a lot about the boll weevil in the Brazos Valley, but for years we didn't recognize that we knew very little about it anyplace else."

Job himself could not have felt as put upon as the cotton growers in the South when the boll weevil burst upon the scene. This was the Cotton Kingdom, and the crop then was uniquely its own. American cotton's entry into international commerce had been a cultural milestone. Cotton, produced from a number of closely related tropical plants, had been known since prehistoric times. Although wealthy Europeans made use of silk, and wool, linen, and flax were generally available, most cotton products until the nineteenth century had to be imported to the West from India, where its manufacture was a high art. The Spaniards found the New World Indians cultivating the plant throughout much of tropical America. The colonists began to grow cotton at Jamestown, but in America it remained chiefly a backyard crop with little promise as a commodity in international trade. Because the downy white fibers adhere to the seeds packed into the boll, removing them from the seeds by hand

is sheer drudgery, a time-consuming and usually unprofit-
able process.

The cotton gin, which Eli Whitney invented in 1793,
changed all that for Americans. (India had developed a simple roller gin built of teakwood long before.) With the related
new technology, it became possible to remove the lint from
the seed and quickly turn it into cotton cloth. Where a
sheepherder could produce perhaps 50 pounds of wool on an
acre, a cotton farmer was able to grow 500 pounds of fiber.
Cotton began to fuel the mills of the Industrial Revolution,
making available to people of all classes a comparatively inexpensive cloth for a thousand uses.

Textile mills might rise in any of the new industrial nations, but throughout North America and Western Europe,
there was only one region suitable for growing this heat- and
moisture-loving plant on a large scale. The Old South held a
monopoly not only on the proper soils and climate, but also,
in an era prior to the development of modern farm machinery, an adequate supply of field labor. The South built
its Cotton Kingdom on its "peculiar institution," and the
plantation system with its complement of slaves advanced
westward through the Gulf states to Texas. When the plantation system was shattered by the Civil War, cotton survived and in the still rural Deep South was the only cash
crop that the well-to-do, as well as the poor whites and
blacks, could depend on. Cotton was not a perishable crop,
as were most foodstuffs in those days before efficient refrig-
eration, nor was it subject to invasions of pest insects that
often decimated crops in other parts of the world. There
were minor infestations from time to time. The cotton leaf-
worm invaded the United States from the tropics and did
some damage, and there were sporadic attacks by the bollworm (*Heliothis zea*) and the tobacco budworm (*Heliothis
virescens*), but these were simply the customary irritations
of a farmer's life.

Then in 1892, the boll weevil appeared in cotton fields along the Rio Grande. There are some entomologists today who believe the invasion was not as abrupt as historians make it out to be, for the weevil sometimes feeds on other plants and it is possible that it lived in the wild on the northern side of the border before it came to the farmers' attention. But cotton is its preferred host. The weevil had made do for ages on wild varieties and scattered cultivated plots of cotton in the American tropics, but stumbing upon Texas it found itself at the gates of the promised land. There was an unlimited source of food, of lush varieties developed by the best agronomists of the day, and no natural enemies to seriously trouble such a tough and adaptable beast.

Homely and earthbound as it may appear at first glance, the boll weevil is extremely mobile when in a traveling mood, and researchers have found that it can fly as far as 45 miles from its point of origin. It swiftly moved out of the Rio Grande Valley and into other cotton-growing regions of Texas. By 1898, it had reached College Station, about 100 miles northwest of Houston, prompting early investigations by entomologists at what was then the Agricultural and Mechanical College of Texas. Other weevils moved eastward, crossing the Mississippi River by 1910 and ransacking the rich cotton lands of the Delta, finally reaching Virginia in 1922. Completely unprepared for a pest of this virulence, the cotton industry—and the communities that depended on it—feared financial collapse. In some areas of the South, cotton was abandoned altogether.

"Domestic cotton by its very nature turned out to be vulnerable to the boll weevil," Knox Walker said. "It takes so long—weeks and weeks and weeks—to mature the fruiting bodies that produce the quantity of lint that's needed to grow cotton economically. The boll weevil is geared to take advantage of the opportunity provided by this extended period. It builds itself up to bigger and bigger numbers, and eventually there are so damn many weevils the farmer is swamped. One year we marked 20 females and put them on an isolated acre of cotton. In the second generation that acre held 5,000 individuals!"

Congress and state legislatures responded to the calls for help from cotton farmers. The agricultural agencies received the funds to hire entomologists, build laboratories, and experiment with the primitive insecticides of the day. Two of the most common were Paris green and London purple, marketed as dyes but found to be also useful against insects because they contained arsenic. There are photographs from the old days that tug at the heart, capturing a sense of desperate optimism, of mules pulling boards on which were balanced partly opened bags of the arsenicals between rows of cotton plants, with their contents gently being shaken down on the leaves. But for the most part, these poisons were ineffective against the resilient weevil. The United States, which had no chemical industry to speak of and relied on German imports until World War I cut off the supply, found itself with no more potent weapons in its arsenal.

"Cotton raising became a race against the boll weevil's capacity to reproduce itself," Walker said. "In humid areas like the Delta and the Southeast, where the plants take longer to fruit, a lot of farmers were put out of business. In this drouthy old God-forsaken country around here, the farmers did a little better."

A new breed of entomologist was coming out of the universities, dedicated to solving the problems created by the boll weevil. Some of them spent their entire careers in the USDA or in state agencies working on this single insect. They devised two strategies, both of which may be classed under the heading of "cultural control." One was to develop fast-fruiting varieties of cotton, giving an edge to the plant in its race with the fecund weevil. The other took advantage of the only weak point entomologists could detect in the weevil's life cycle. The season's last generations, heading toward diapause and eventual re-emergence the following spring, required a source of food in late summer and fall. Generally they found it on what was left of the plants after farmers had completed their harvest. Entomologists suggested destroying the stalks immediately after harvest, thus depriving the weevils of the fat accumulations needed to carry them through the winter.

Some farmers took more drastic steps and moved away from the warmer, humid regions preferred by the weevils and thus evaded the problem entirely. Tennessee, the "Boot Heel" of Missouri, and high plains of northwestern Texas proved to be sanctuaries from the boll weevil's depredations. And, in time, new cotton-growing regions were opened farther west in Arizona, New Mexico, and California. Other farmers found some relief in a new insecticide from the American chemical industry that had been developed during World War I. Calcium arsenate, applied to fields by the new technique of aerial crop dusting during the 1920s and 1930s, offered a promise it never quite fulfilled. However, techniques for application were then worked out that later came into general use in a more chemical-oriented age. So were sales techniques, as chemical companies learned to peddle their panaceas to worried farmers. (One new insecticide formulation of the time was marketed under the brand name Genicide.)

"Some cotton farmers in Texas bought these chemicals before World War II, but a lot more of them didn't," Knox Walker said. "Either way, it was no picnic. For poor whites and poor blacks in large areas of the South, cotton was the only cash crop they had. They went on living alongside the boll weevil, and they would grin and bear it."

World War II brought with it a new technology and American agriculture rushed headlong into its euphoric years. The development of DDT and related chlorinated hydrocarbon insecticides seemed to provide growers with the ultimate weapon against insects, prompting them to abandon the careful cultural practices built up over decades of painful experience. But in the end, cotton growers in Texas were to be confronted by the same disasters that overtook their counterparts in California's citrus orchards and in New England's forests. In fact, the most lurid case histories of the new chemicals' failures were to come from the cotton industry.

"The boll weevil is the centerpiece in Texas," Knox Walker said. "It's what makes you do the kind of things that get you in trouble with the bollworm or the tobacco budworm or any of those other pest insects."

The urge to use chemicals against any insect that showed itself in the field became irresistible. Its apparent origin was in sales campaigns mounted by the chemical companies, although the message was reinforced by entomologists in the USDA, the state extension services, and the big land grant universities.

"In my opinion, the chemical companies and their message have changed," Walker said. "In the years after World War II, they were spreading the word that it was an American virtue, like eating black-eyed peas, to use chemicals. We had defeated the Axis Powers, and now we had DDT, and we were on top of the world. In retrospect it was very shortsighted, but if you were the head of this Entomology Department at the time, you'd probably have had your people doing the exact same thing. There was no precedent for this so that we could figure out the pitfalls. We hadn't even produced many chemicals of any kind until World War I. We just didn't have any common sense about them."

But the message was blared to a receptive audience through radio commercials, billboards, and agricultural publications. The Hercules Powder Company, manufacturers of insecticides, urged farmers to adopt its "washday policy"— spray the fields each Monday. Never mind the level of infestation (or even its absence); a blanket of insecticides was good insurance. The proponents of this approach urged that for the nation to let down its chemical guard would be to expose itself to hunger and disease. But the truth was that the lion's share of insecticides goes not to food crops or public health projects, but to cotton. (It is estimated today that approximately 47 to 49 percent of the volume of insecticides used on crops in the United States is used on cotton.)

Regions around the world in which no one would have dreamed of growing cotton a few years earlier were suddenly opened up to intensive cultivation. Northeastern Mexico and northwestern Australia experienced a few years of high cotton yields. "Cotton never should have been grown in those places because the insect problems were too intense," Walker said. Long-fruiting plants and constant applications of chemicals made it profitable at first. Then reality asserted itself. The insects became resistant to the chemicals, new pests appeared, costs for the required new chemicals skyrocketed, and the schemes ended in financial collapse.

One of the areas hardest hit in the late 1950s was the Rio Grande Valley. Scientists had developed strains of long-fruiting cotton that took 180, and perhaps 200 days to mature, with a concomitant bumper yield of fiber. Production more than doubled on some farms to better than two bales an acre. The traditional race between the maturation of the plant and the build up of the boll weevil population was no longer a consideration. Repeated applications of insecticides bombed the weevil off the scene.

When boll weevils became resistant to the chemicals,

around 1954, there was a reluctance to accept the evidence. Podsnappery remained the prevailing mood. John H. Perkins, in his book *Insects, Experts, and the Insecticide Crisis,* writes that Ky Pepper Ewing, chief of research on cotton insects at the USDA, fought the evidence of weevil resistance uncovered by researchers Dan Clower and John S. Roussel under L. Dale Newsom at Louisiana State University.

"He tried to persuade Dale Newsom not to release LSU's data gathered in 1954–1955 until more information was available," Perkins wrote. "Ewing was aware that resistance had been shown in other insects, but he feared that release of Clower and Roussel's data would damage both the insecticide and cotton-growing industries. Newsom recalls that Ewing was reluctant to accept the idea of genetic heterogeneity within boll weevil populations and may not have accepted the theory of organic evolution. Ewing was an evangelical Methodist, and perhaps he held fundamentalist beliefs contrary to evolution."

As evidence of resistance overwhelmed all doubt, the agricultural community changed weapons. The DDT was mixed with toxaphene to achieve a synergistic action, which was effective for a time. But the boll weevil always adapted, and one by one the chlorinated hydrocarbons were discarded. The experts suggested parathion and later methylparathion, keeping the weevil on a tight rein if not vanquishing it by the sheer volume of chemicals. But the insect world is a Hydra. When one head is lopped off, two more appear in its place, and this was what happened in the cotton fields. Other insects that had never been considered consistently damaging now emerged as major pests. Those destructive twins in the genus *Heliothus,* the bollworm and the tobacco budworm, came on like tigers. The pink bollworm, an exotic insect long known as a pest in India, lodged in American cotton fields with a tenacity reminiscent of the boll weevil's. Cotton growers in the Rio Grande Valley and some other areas in the South were all but wiped out.

Yet almost every element in the community seemed to conspire to keep the farmer on chemicals. The salesman and the agronomist were in agreement. The Farm Home Administration even attempted to include in its loan agreements, in certain cotton-growing areas, a stipulation that the mortgagee would follow the rigid schedules of chemical application suggested by local agricultural experts. Throughout most of the South, there seemed to be no alternative to the constant rain of insecticides.

18

THE
COTTON CONNECTION
The Teamwork Strategy of Integrated Pest Management

COTTON, LIKE MOST OTHER PLANTS OF ECONOMIC OR ESTHETIC value to mankind, is constantly in flux. Its size, the shape of its parts, even its component chemicals, are altered by agronomists to meet the demands of the moment. There was an extensive gene pool to begin with, for at least 30 species of cotton plants are known from the wild in various parts of the world. Of those, only four have been selected for cultivation to any extent. The crop grown in most of the United States, although it appears in many varieties, derives from a New World species; it is called upland cotton (*Gossypium hirsutum*).

Cotton requires a long growing season, warm temperatures, and moist, fertile soil. The southeastern United States, where the industry has flourished for nearly two centuries, provides these requirements and consequently yields a luxuriant, high-yield crop under natural conditions. Farther west, where the industry spread in later years, irrigation is required to provide the needed moisture. The lush, long-fruiting varieties may stand five feet or more tall, the short-season varieties no more than two feet. Although the yield per acre, depending on moisture, fertilizers, and plant varieties, is generally between one and two bales (a bale contains about 500 pounds of cotton), the yield in certain areas of California's rich Imperial Valley is three to five bales an acre.

Besides cotton's lint, cottonseeds yield a useful oil. Thus, cooking oils and oleomargarine are by-products of the ginning process, as is cottonseed cake for cattle feed. Cottonseed products for consumption by humans or animals must be diluted, however, because the plant contains the mildly toxic substance gossypol. The floral parts of the plants also produce nectar, which attracts pests, beneficial insects such as parasitic wasps, and predatory ants.

No matter how strong one's attachment to the chemical approach to pest control, there had to be serious questions about the future of cotton growing in Texas, where nearly one-third of the nation's crop was grown. By the late 1950s, the industry was in a shambles in the Rio Grande Valley. In most other areas, the growers were holding on, getting high yields from their long-fruiting plants, but at a constantly rising price for insecticides, fertilizers, and water. Agricultural officials now saw the need for intensive research on the entire complex of cotton problems and the development of more effective insect control programs. The State Experiment Station brought in new talent to Texas A&M, including the highly regarded entomologist Perry Lee Adkisson.

Adkisson had received his early training at the University of Arkansas. His teacher there was Dwight Isely, an expert on the boll weevil, whose experience had led him to believe that insecticides should be used against a pest only as a last resort. He had conceived the idea of "scouting" fields to determine the levels of infestation. Adkisson served as a scout in Isely's program at Arkansas, checking fields for signs of a serious weevil outbreak so that growers would not have to bear the expense of chemicals until they were actually needed. After further study at the University of Kansas, Adkisson found a position in Missouri. The call to Texas came in 1958.

"The pink bollworm, which I was to work on in Texas, was already on the way out as a major pest when I arrived," said Adkisson, a small energetic man who is now Deputy Chancellor for Agriculture at Texas A&M. "There had been the beginnings of a return to sound cultural practices like the destruction of stalks right after harvest, and this had helped with the pink bollworm. As we knew from the past, stalk destruction was also effective against the boll weevil and we started to build on that. But the philosophy in Texas still leaned toward the automatic treatment schedule—what Dale Newsom at LSU called a 'womb to tomb schedule'— and a lot of the chemical company people advocated it through various farm organizations. But some of us saw what was happening. We began to question the season-long treatment and we pushed to get scouts into the field."

The attempt by Adkisson and his colleagues to convince cotton farmers to cut down on the use of costly insecticides ran into determined opposition. Chemical companies, caught up in the postwar euphoria and forecasts of greatly increased sales, had invested heavily in plant expansion. A

Cotton flower, boll, and opening boll.

whole generation of farmers, entomologists, and agricultural administrators had come to maturity in an era dominated by the belief that the new chemicals would provide the final solution to pest problems. But Adkisson advanced rapidly in Texas. He was put in charge of the Cotton Insect Laboratory, and then in 1967 he became head of the Entomology Department at Texas A&M. He now had a strong voice in the direction pest control would take in Texas and he was interested in that mix of control techniques that became known as integrated pest management.

"We decided to get more into plant resistance to insects," Adkisson recalled. "Also, we knew we were using too much water and too much nitrogen fertilizer. We had been growing a big luscious stalk, highly attractive to the *Heliothus* insects. We would be reducing yields, but we would also be reducing costs."

As Adkisson and his team slowly began to win over the agricultural community in Texas, he found the major challenge to his integrated pest management approach in the USDA. Ironically, the leader of a second major—and competing—pest control strategy was a scientist who had already embraced one of those "other roads" advocated by Rachel Carson in *Silent Spring*, Edward F. Knipling. Elated by his success against the screwworm, Knipling conceived a grand plan to eradicate the boll weevil from the United States. He seems to have viewed his plan much in the nature of integrated pest management, using a variety of techniques to reduce the weevil population to very low numbers and then applying the coup de grace with his favorite weapon, the sterile male.

Adkisson admired Knipling and for a while was sufficiently interested in his plan to serve on the Technical Guidance Committee responsible for setting general policy on an experiment to test the plan's feasibility. The idea of eradication must appeal to anyone who has fought the boll weevil for many years.

"The sterile male technique is absolutely essential for eradication," Adkisson has said. "You've got to have a lot of males out there searching for that last fertile female."

But the eradication experiment, which centered on a cotton-growing area in southern Mississippi, caused Adkisson to change his mind about the plan. There were serious flaws in the operation. The tough boll weevil resisted sterilization in the rearing facility and fertile males were released along with sterile males. Any such plan depends extensively on the full cooperation of all the growers; but in Mississippi, some who had planted cotton simply to collect federal price supports did not bother to spray their fields or take the other precautions needed to bring the weevil population down to low levels. Moreover, it was soon apparent that what reducing the weevil's population to extremely low levels really meant was blanketing all the cotton-growing regions of the South with enormous amounts of insecticides. Having observed the eradication experiment in Mississippi (and a later one in North Carolina), Adkisson withdrew from the program.

"I believe that eventually we could have the technology to make boll weevil eradication feasible, but I do not believe it could be economically feasible," Adkisson said. "The growers and the states are asked to pay 75 percent of the costs of the program, which would run 70 to 80 dollars an acre. That may be cost-effective in North Carolina, but it would not be cost-effective in Texas. Under our present control techniques, that's more than some of our farmers would spend on insecticides for boll weevil control in 15 or 20 years."

Other critics cited the terrible environmental costs of such a program. Knox Walker, pointing to the boll weevil's mobility and resilience, just shook his head and chuckled.

"I can't imagine how they would make eradication feasible over millions of acres," he said. "You'd be fighting an endless battle of reinfestation. If you enjoyed the Vietnam War, you'd enjoy working on the boll weevil eradication program."

With Knipling's retirement from the USDA, the dream of eradication is fading, although diehards in the Southeast fight on. Perry Adkisson and his team in Texas are trying to prove that a many-pronged approach, working within the boundaries established by nature, is the way of the future.

"Some people say that integrated pest management is just an ivory tower project," Adkisson said in 1982. "But go to almost any cotton farm in Texas and you will see that the farmer is using its principles."

Texas stands out from most other states in the cotton belt. Elsewhere, integrated pest management may be simply a fashionable term to describe control programs that employ field scouts chiefly to give growers the signal when to fire the big chemical guns. Even at Texas A&M, individual scientists have reservations about certain aspects of the integrated program. There are differences of opinion about the specific worth of this or that natural enemy, or this or that genetic trait nurtured to produce "weevil-proof" strains of cotton. For instance, several entomologists there shrug off a new strain called Frego cotton, whose twisted bracts do not close over the boll and thus tend to discourage egg-laying weevils; its critics point out that the distorted boll leaves the plant vulnerable to other pests. But in general, there is unusual cooperation in devising the separate elements that make up an integrated program.

"On many land grant campuses there is a wide gap between the research and the extension service people," Adkisson said. "The researchers seem to feel that their responsibilities end when they publish their research. But here the scientists work with the extension specialists in the field, seeing that their work is implemented."

What Adkisson refers to as the "rationale" for the short-season approach to cotton production is based on work carried out by Knox Walker and the agronomist George A. Niles at Texas A&M some years ago. These scientists showed that, within a month of flowering, the short-season variety began to set bolls almost invulnerable to weevil oviposition after only 12 days of growth. Moreover, these tough, early-season bolls are the ones that tend to produce the bulk of the cotton crop. Thus, the crop is vulnerable only when the small, overwintering generation of weevils is in the field, quickly becoming secure before the more abundant generations emerge. (The long-season varieties are just setting their bolls as the weevil's numbers build to pest proportions.)

Luther Bird, a heavy-set man with a deep drawl, is Professor of Plant Pathology at Texas A&M. For two decades he has worked to develop cotton varieties that will stand up to the many hazards the plant is heir to. Several of them, bearing designations such as TAMCOT SP-21 and TAMCOT SP-37 (the letters, of course, stand for Texas A&M cotton), were ready for commercial distribution in the early 1970s, just as the integrated program was swinging into gear. Besides fruiting rapidly, the new varieties possessed genetic traits that enabled them to tolerate cold and to resist bacterial blight. Growers consequently found them suitable for the early planting required to get the jump on the boll weevil.

Bird concentrates on the interrelationships among genes that cause resistance to various diseases found in cotton. He has referred to the results by the imposing name multi-adversity resistance system (MAR).

"Cotton, like other plants, has its own natural symbiotic flora—minute organisms such as bacteria that grow right along with it," Bird said. "These organisms grow in the plant's tissues, as well as on the surface of its roots and seeds. Well, the basic idea of MAR is that cotton has the genetic potential to alter its microflora so that it acquires a complement of organisms that are inherently hostile to the insects and pathogens that normally attack the plant. The process is under the control of what we call MAR genes. These genes alter the quality and quantity of the constituents of fluids in the cotton plant's tissues or the exudates of the seed coat and roots. The altered fluids are nutritionally unfavorable for pathogens and insects, but selectively favorable for fungi and bacteria that are highly competitive with those harmful organisms. So, to sum it up: Unfavorable nutrition and microorganisms work together to provide a mechanism for multi-adversity resistance."

Bird fosters resistance against bacterial blight in cotton by selecting seed from varieties that have other favorable characteristics. From them, he grows seedlings in the laboratory in soil infested with seedling pathogens. When they are about five days old, he inoculates the seedlings by scratching them with a toothpick he has dipped in an inoculum composed of four races of the bacterial blight. If, after 10 days, the seedlings show no blackening indicative of any of these diseases, he transplants them to larger pots and cultivates them as a promising new variety, with MAR resistance.

Although Bird was not aware of it at the time, the new plants possessed resistance to some insect pests as well as to cold and disease. The bacteria and other microscopic organisms associated with the plant tissue proved to be unpalatable to those insects and deterred them from feeding on the plants. These varieties, released for sale by seed companies, have been credited with saving the cotton crop in the Coastal Bend region around Corpus Christi, where the growers had fallen on such hard times that many of the gins had been shut down.

"Oh, the farmers fight over these seeds when they first become available," Bird said. "I've had companies tell me they have to put a guard over the seed. Like a farmer down there says, 'At least now I look my banker in the eye and grin. I never could do that before.' "

Bird tests the new varieties on working farms with which he has close contacts in several counties throughout the state.

"In fact, I put more confidence in some of these tests we run on the farms than I do in our more formal experiments," he said. "There's lots of subtle things in this resistance business. Some of the insect resistance we knew about first from the farmers. We had these new varieties out there in the fields with other cottons, and the farmers told me they'd get out their little old notebook, where they'd be checking through with their thumb, and they'd say, 'You know, the boll weevil doesn't like this one as much as it likes that one.' That's the way you get your leads. And then you go back and you set up more experiments and you pin it down."

Bird continues to develop certain genetic traits from the great variety of cotton plants maintained at Texas A&M. The typical cultivated plant bears hairy leaves. The bollworm and tobacco budworm lay their eggs among those hairs. Other scientists discovered that the eggs are less likely to be laid on hairless plants, and for this reason, Bird is developing hairless varieties of cotton. He is also at work on a glandless variety.

"These are the glands that produce gossypol, the toxic substance in cotton plants," Bird said. "It used to be that we wanted to keep the gossypol in there because it was toxic to some insects. But that's been broken now. These new glandless varieties have more resistance to insects than some plants with glands. It makes cotton seeds much more valuable as livestock feed. It also opens up whole new opportunities to develop cotton meal for human consumption. Cotton is really a much more palatable protein than soybeans."

Luther Bird put the problem in perspective for the Texas cotton grower. "These new plants will help him," he said. "With the old varieties, the farmer had to help *them* all the way."

Despite the integrated program's capacity to take advantage of natural enemies, imaginative cultural practices, and resistant varieties of cotton, there are conditions under which the boll weevil and the plant's complement of other pests escapes control. "We still need insecticides in Texas," Knox Walker said. "It's naive to think we can get along entirely without them, because farmers are growing cotton in some places that really aren't suited for it. Environmental conditions, such as favorable temperatures and excessive moisture, create enormous numbers of pests, as they do in the Rio Grande Valley. High production is important there, so that they use long-season cotton and loads of chemicals. Natural enemies don't stand a chance."

But such high production areas account for only about 150,000 of the 7 million acres on which cotton is grown in Texas. In most places, then, natural enemies are vital in producing a profitable crop. Any application of chemicals threatens to destroy those valuable allies, thus creating pests from previously innocuous insects. But when, for some reason, pest insects slip out from under natural controls, the grower understandably takes energetic steps to save his crop, in that case resorting to a quick fix with chemicals.

Natural enemies disappear and he finds himself saddled indefinitely with chemical costs.

Why should insecticides almost invariably take a greater toll of natural enemies than of pest insects? Predators are especially hard hit. A large percentage is killed by the insecticide itself, while secondary factors compound the damage. Predatory insects are less numerous than plant-eating species (just as foxes are less numerous than mice and rabbits). This lowers their chances for survival, and when their food (the pest species) becomes scarce after chemical treatment, predators are adapted, through evolution, to adjust by bearing fewer young per litter, which in turn lowers both their potential for increase and the odds of producing a resistant variant. Finally, by consuming large numbers of already poisoned pest insects, predators that survive the direct attack accumulate massive doses of the compound and eventually die. Similarly, parasites have the statistical deck stacked against them. Many that survive the initial insecticide application eventually perish because their host has suffered a direct hit. In any event, the naturally prolific plant-eating pests snap back quickly in the absence of an effective population of natural enemies.

In the past, the agricultural community has glibly suggested that this problem can be corrected by the use of "selective insecticides." The term has an ecological ring to it, but in most cases (as we have just seen), a poison powerful enough to kill pest insects does an even more effective job on their natural enemies. In recent years, however, Frederick W. Plapp, Jr., Professor of Entomology at Texas A&M, has been able to single out insecticides that do not load the odds so heavily against natural enemies. The difficulty here is that most insecticides are inherently more toxic to predators and parasites than to plant-eaters.

"The relative toxicities of insecticides to different in-

sects are functions of the ways insects detoxify them," Plapp
has explained. "They detoxify most poisons by one of two
routes, hydrolysis or oxidation."

All insects, natural enemies as well as pests, can detoxify substances by using their hydrolytic enzymes. That is,
they can decompose chemical compounds through a reaction with water, putting to use the enzymes whose normal
function is to metabolize fats and proteins. A major obstacle
in programs that try to use both chemicals and natural enemies against pests is that most of the insecticides used in
recent years cannot be detoxified, or metabolized, hydrolytically.

"Here the pests have an important advantage," Plapp
said. "Most pests are plant-eaters, and only plant-feeding insects have efficient oxidative detoxifying systems.
In fact, they must have these enzymes to render harmless to themselves the toxic substances, such as gossypol, that are often found in plants. Since these enzymes
also detoxify insecticides that are subject to oxidative
metabolism, plant-feeders have a natural tolerance to
these chemicals that is absent in predators and parasites. The idea, then, is that by using insecticides that
can be detoxified hydrolytically, we may be able to
achieve chemical control without completely eradicat-
ing beneficial insects."

In experiments with the tobacco budworm and its parasites and predators, Plapp worked out tables of selectivity,
rating the level of toxicity of various insecticides to various
insects. Parathion and the other organophosphate insecticides, as expected, proved to be much more toxic to beneficial insects than to plant-feeders. On the other hand, some of
the newly developed pyrethroid insecticides coming into

more common use on cotton are better able to be detoxified by predators and parasites. Pyrethroids, then, may be the choice in the future, when experts in integrated pest management must resort to chemicals.

But for some of the scientists at Texas A&M, the younger ones especially, the ideal future for the control of cotton pests would be free of chemicals, selective or otherwise. In the next chapter we shall examine the state of their art.

19

THE
COTTON CONSERVATORS
*The Integrated Pest Management
Success in Texas*

CARDIOCHILES NIGRICEPS, A BRACONID WASP ABOUT THE SIZE OF a housefly, is common in Texas cotton fields. It parasitizes both pestiferous members of the genus *Heliothis,* the bollworm and the tobacco budworm, but is effective only against the latter. This little wasp, then, furnishes an instructive example of the delicate balance that exists between a parasitoid and its host.

Most beneficial insects, once they have arrived in a habitat suitable for their hosts, apparently locate their targets somewhat at random. *Cardiochiles* functions more as a bloodhound, picking up a chemical trail and following it doggedly to its victim. Entomologists call these chemical clues kairomones. The more familiar term pheromone refers to the chemical that an insect secretes as a signal to other members of its own species; a kairomone is a chemical that an insect secretes as a signal to individuals of another species, ironically, to the benefit of the second species. In this case, the kairomone is a mixture of methyl esters produced by a gland in the mandibles of the tobacco budworm.

Cardiochiles, following this chemical trail, unerringly locates its host. After ovipositing, the wasp secretes a chemical of its own to mark the host as well as the surrounding area. This cunning precaution reduces the chances that another female will deposit *her* egg there, and thus helps to spread the new generation of parasites evenly throughout the field. The chemical also advises the original wasp that she has already searched that particular area and need not visit it again. The single egg she has laid is sufficient to bring the tobacco budworm to grief.

Yet if *Cardiochiles* deposits its egg in a bollworm, the egg will not develop, although some parasites are extremely effective in controlling the bollworm. *Cardiochiles,* however, fails because the bollworm's tissues immediately respond by encapsulating the egg.

A student at Texas A&M who was observing parasitism in the tobacco budworm several years ago noticed the presence of strange looking particles around a parasite's egg. He called them to the attention of his professor, S. Bradleigh Vinson. Vinson recognized the particles as viruses. He and his students thus began an investigation that has increased our knowledge of the complex relationship between an internal parasite and its host.

They discovered that many parasites harbor a virus in their reproductive systems, which is injected into the host along with the eggs. Without the virus, the parasite's egg is unable to defend itself against encapsulation. The virus not only inhibits the host's immune mechanism in some way that is still unexplained, it also causes nutrients to remain in the host's bloodstream, where the developing parasite can extract them. Only with the aid of the virus can the parasite compete with the host's tissues for food. This may be the first recorded case, according to Vinson, of a virus-animal symbiosis.

A number of experiments followed from this initial discovery. When researchers injected the virus alone into the host, the host responded as if it had been effectively parasitized. It no longer fed or developed normally, and soon died—not as if it had been attacked by a virus, but as if it had been drained by a parasite.

Going a step further, the researchers extracted a virus from a species that survives as a parasite in the bollworm. They injected the virus into a bollworm that had just been parasitized by *Cardiochiles,* which, as we have seen, seldom survives in that host. This time *Cardiochiles* flourished. When it emerged as a healthy adult, the wasp carried in its own body the life-giving virus.

Genetic engineering may permit scientists in the future to implant other viruses in the *Cardiochiles* population as a whole, transforming it into an effective, all-around natural enemy of the genus *Heliothis* in cotton fields.

There is a Texas prison farm in fertile bottomland along the Trinity River, north of Huntsville. Its formal name is the Ellis Unit of the Texas Department of Corrections, but most people know it as "the Ellis Farm," and cotton has been grown there successfully for many years.

"By the time I went over there as a researcher, the state had reformed its prison system," Winfield Sterling said, "and there was nothing like the conditions you used to see pictures of, with prisoners working in the fields with chains around their necks. They were still harvesting cotton by hand, but in the last few years even that has changed and now machines perform a lot of the work."

Sterling, a Professor of Entomology at Texas A&M, is a trim, articulate man with reddish hair and glasses. His father raised cotton, as well as nine children, in the Rio Grande Valley, and Sterling went to school down there at Pan American University. Naturally, he knew a lot about cotton and its complex of pests long before he completed his graduate work at Texas A&M in 1969. Having seen for himself that the *Heliothis* worms could not be controlled with chemicals in South Texas and the Coastal Bend region, he was open to new ideas about insect control when he joined Perry Adkisson's team.

"Growing cotton had become a losing proposition over at the Ellis Farm, too," Sterling said. "*Heliothis* was a major problem for them. Their insecticide costs had tripled while their yields had been cut in half, and the Department of Corrections was considering getting out of the cotton business. They asked for help and I was part of the group—it included Perry Adkisson and Knox Walker—that set up a demonstration project there in 1968. I've continued with field research at the Ellis Farm ever since."

Winfield Sterling is one of the scientists whose research has helped to put together a remarkable program of integrated pest management against cotton pests in Texas. Some of the work has been a resounding success and is now standard practice in Texas; some of the work remains theoretical (although highly exciting) or controversial; and some of the work was discarded when it led into blind alleys. No lead has been neglected. Sterling, for instance, has raised eyebrows all over the South because of his insistence that not even the much-despised red imported fire ant (*Solenopsis invicta*) should be neglected in the pursuit of tools with which to combat cotton pests.

"A big handicap we have in entomology is that we want to put everything in black and white," Sterling said. "An insect has to be all bad or all good, we think. But that's not the way it works in nature. My concept of pest management is working almost on a field to field basis, using the native natural enemies as much as possible."

The red imported fire ant (that is, imported inadvertently to the United States from South America) stands branded in the public eye as a monumental pest itself. There was some evidence to support the notion, however, that this notoriety stems more from the vivid imaginations and matching prose of the USDA's public information specialists than from the real-life threat of fire ants to crops or livestock. Because their sting justifiably prejudices people against them, the fire ants presented empire-building administrators within the USDA with an irresistible target in the 1950s. The agency's massive eradication campaign, launched at a tremendous cost in both public funds and wildlife, became the greatest fiasco in pest control history and probably helped the fire ants to increase their range throughout much of the South. On the other hand, the fire ants' good name (they are looked upon as rather helpful predators in their native habitat of Brazil) never recovered from the uproar.

Sterling and his colleagues realized at once that the Ellis Farm, because of its isolation from private agricultural land, provided an excellent site for both research and for a demonstration project. They selected fields isolated from areas where chemicals were regularly applied to crops, and worked to eliminate drift into the research fields. Although the land was owned by a state agency, the entomologists acted on the assumption that the ultimate test of the experimental methods was to be economic. Would they turn a profit?

"We were trying for a rapidly growing crop, with early senescence, or termination, to keep the boll weevil generation that is headed for diapause from building up to large numbers," Sterling said. "We wanted to harvest a crop in under 150 days."

Working with approximately 1,000 acres of cotton on the Ellis Farm, the researchers were able to monitor the effects of the new integrated methods on what had been a seriously disturbed agricultural system. Repeated applications of organophosphate insecticides had knocked down the population of boll weevils, while permitting the *Heliothis* insects, resistant to those chemicals (as the weevils are to chlorinated hydrocarbons), to emerge as the "key" pests. The entomologists encouraged the planting of rapid-fruiting, short-season cotton on the Ellis Farm. In some fields, no insecticides were used during the growing season, and water and fertilizers were kept to a minimum. When the crop was harvested, small quantities of insecticides were applied to the plants to kill as many late-season boll weevils as possible before they left the fields for their overwintering quarters on the fringes. Later, the entomologists encouraged the workers to shred and plow under the stalks of the old cotton plants to destroy the egg-laying and sheltering places of *Heliothis*. The results were dramatic from the start.

"Budworm-bollworms on the Ellis Farm must be considered to be largely a man-made problem," Sterling and one of his colleagues, Robert L. Haney, wrote in the publication *Texas Agricultural Progress*. "The evidence for this is that no insecticides were used for the control of these pests during 1971 and 1972 while at the same time an increase in yield was obtained. Thus the control costs were reduced from $17.47 per acre in 1969 to zero dollars in 1971 and 1972."

Cotton production on the Ellis Farm was once more on a paying basis. The *Heliothis* twins, bollworm and budworm, fell back under the control of the resurgent population of predatory insects. To determine exactly what was happening in the fields, Sterling and his colleagues labeled the eggs and larvae of the tobacco budworm with radioactive phosphorus and distributed them in one of the fields. Later, using a Geiger counter, they searched the field for predatory insects that had consumed the eggs and larvae of the budworm (as indicated by the traces of radioactivity in the predators' bodies) and brought them back to the laboratory for analysis. There, the researchers placed the predators in a scintillation counter and recorded the counts per minute of the radioactive material, thus calculating the number of budworm eggs or larvae eaten by an individual predator. Spiders of various kinds had devoured a surprisingly large number. There were many other voracious predators out there, as the analysis proved, including pirate bugs, big-eyed bugs, green lacewings, and red imported fire ants.

"For a *Heliothis* egg to be laid out there is suicidal," Sterling said in wonderment. "Look at these results of our studies. Assuming a female lays 1,000 eggs, we found that about 926 are taken by various predators. Then we follow this population of *Heliothis* through each larval instar, and we find that the predation continues unabated so that by the fifth instar the survival rate for the original 1,000 eggs is .01, or 99.99 percent mortality. And we see this year after year."

As *Heliothis* declined in the unsprayed fields, the boll weevils became more numerous. Yet only once in the following decade did boll weevils approach a level of infestation (attacking 15 to 25 percent of the young squares) at which practitioners of integrated pest management in Texas would recommend chemical treatment; and even then the boll weevils did not damage the crop economically. In other years, they infested less than 5 percent of the squares. What was controlling the weevils in the absence of chemicals?

Ever since he was a boy, Sterling had heard that the boll weevil had no serious natural enemies in the United States. One day, while examining plants at the Ellis Farm, he saw fire ants kill a boll weevil. But when he mentioned the incident to a fellow researcher, the man shrugged it off. "Yes, fire ants will take a weevil once in a while, but they're not a serious threat to them," the man assured him.

The authorities at the Ellis Farm did not ordinarily use chemicals against the fire ants in cotton fields, as farmers often do. Those insects posed no serious threat to crops or livestock, nor even to workers in the fields, for fire ants do not usually attack a large creature unless it blunders into the nest. Fire ants, therefore, were abundant on the Ellis Farm and Sterling had ample opportunity to watch them forage. By occasional samplings, he found fire ants in large numbers on cotton. He knew that, as lovers of sweet substances, fire ants would be attracted to the nectar secreted by the plants (an evolutionary ploy designed to lure pollinating insects). The aphids that lived on the plants also attracted them with their honeydew. But the aphid population fluctuated widely, being preyed upon by ladybird beetles, and thus the ants apparently did not find aphids a consistent source of food.

The primary cause of death early in the boll weevil's life cycle in many parts of Texas is desiccation. Especially in very hot weather, anywhere from 50 to 80 percent of a given generation may succumb. Other individuals may not progress beyond the egg stage because of infertility or encapsulation by the boll itself. The dominant hymenopterous

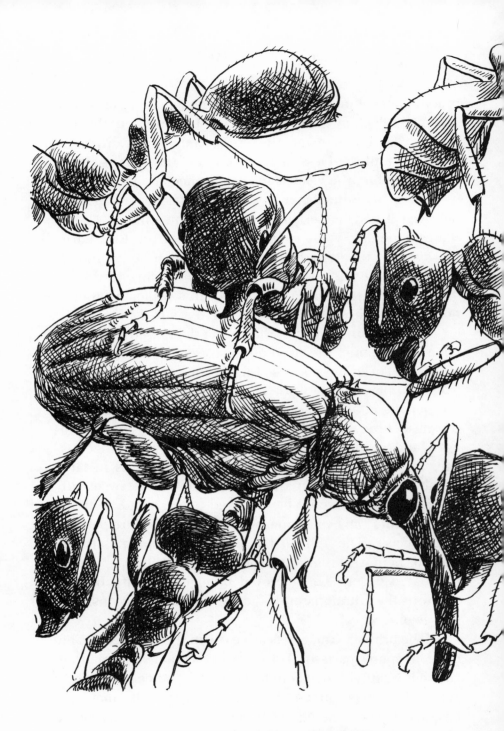

Fire ants are effective enemies of the boll weevil.

parasite that attacks the boll weevil in the square or boll is *Bracon mellitor;* but this insect rarely kills more than 5 or 10 percent of the weevils.

Sterling and his fellow researchers began a series of controlled experiments to discover how effective fire ants might be in preying on the early stages of the boll weevil. They selected one field in which fire ants flourished and another in which they destroyed the ant population by the use of Mirex baits, which are taken only by ants. Later, they searched the ant-abundant field for cotton squares that contained the obvious holes in their walls indicating that some creature had passed either in or out.

"Squares with such holes were recorded and brought back to the lab for determination of the source of the hole," the researchers wrote afterward in *Environmental Entomology.* "If, upon square dissection, a weevil pupal skin and white fecal pellets were found inside a well-formed pupal cell, the hole was attributed to an emerging adult weevil. If a silken cocoon, empty or occupied, along with a dead immature weevil was found, weevil mortality was attributed to a weevil parasite. No naked parasite pupae or puparia were observed during the study. If the square contained one or more characteristic ant predation holes and inside there was no trace of a weevil or parasite cocoon, the hole and weevil mortality were attributed to ant predation."

Because earlier studies had shown that scavenging ants do not usually open squares containing weevils already killed by heat, the researchers concluded that their results provided an accurate picture of ant predation. However, their studies established that this predation was ineffective against boll weevils on cotton plants. Thus, predation on the plant was not high enough to explain the difference between the low level of weevils in the field where ants flourished and the high level of weevils in the control plot in which the ants had been poisoned. Cotton squares containing weevils eventually turn yellow and drop from the plant. Was it possible that most of the fire ants waited for the square to drop before entering them? The researchers turned their attention to squares on the ground beneath the plants.

"We found that the ants had killed the weevils in almost 100 percent of the fallen squares," Sterling reported. "Soon after the squares landed on the ground, the fire ants attacked them and cleaned up the larvae inside. Often we actually saw the ants preying on the weevils. Later, the ants would hit the young adult weevils that were drilling a hole to get out of the square. They even went into the green bolls after them just when the boll began to split and the weevil was very exposed and about to emerge."

The researchers had calculated that if predators took 80 percent of the young weevils—the survivors after desiccation, encapsulation, and other natural factors had begun the culling process—the population would be reduced to about 2 percent of the original total. This level of predation guaranteed that the weevils would not pose a threat to the crop during many years. Obviously, fire ants alone were providing such protection on the Ellis Farm.

"The only damage to the crop was coming from the hard-shelled weevils that had overwintered on the edges of the fields and were too hard for the ants to bother with," Sterling said. "It gave me a lot of confidence in our program to see that fire ants were not only maintaining low levels of boll weevils without the use of chemicals, but even when the weevils got a big head start, the ants were able to catch up with them."

Sterling has not advocated the red imported fire ants as a cure-all for boll weevil problems in Texas. They are not effective in cooler regions to the north. Nor are they effective even in the heart of the range in late summer when the weather cools and the final generations of weevils head into diapause; temperature seems to be closely linked to the fire ants' effectiveness as predators. But surveys made by Texas A&M students in other areas of the state prove that, wherever fire ants are abundant and chemicals are restricted, predation rates are high. In areas where the red imported fire ant is absent, native ants provide farmers with some protection from the boll weevil. Well aware of regional sensibilities, Sterling would not dare recommend propagating fire ants for boll weevil control. All he asks is that cotton growers who already have them on their property begin to consider fire ants from a new perspective.

"I call myself an insect ecologist," Winfield Sterling said. "I would not say it's a *mistake* to spray, because in some situations it is undeniably profitable. But environmentally, spraying is harmful. Our program at the Ellis Farm has shown that a grower can often refrain from using chemicals and still produce short-season cotton with a yield at least as high as the state average. In the South, that statement is heresy."

Some of the best research scientists at Texas A&M, like those at any great institution, keep their eyes on the future. Among them is S. Bradleigh Vinson, who completed his graduate work at Mississippi State University, concentrating on the phenomenon of DDT resistance in boll weevils.

"But I am a physiologist, interested in insects rather than insecticides," Vinson said in a comparatively free moment before rushing off to the nearby USDA laboratory where he maintains the bulk of his experimental insects. "When I got my degree, I just didn't feel that insecticides were the wave of the future. I had become fascinated with parasites."

The reason for his shifting interest was the work of a

British scientist, long since retired, named George Salt. Salt had tried to find out exactly how parasites locate their hosts.

"He had a lot of good ideas but he didn't have any background in chemistry or physiology so he wasn't able to follow them up," Vinson said. "I have a chemistry background so I've been able to take some of his ideas a little further. That's the kind of thing I've been doing here."

What does a parasite "cue in on" when it enters a field? *Cardiochiles nigriceps,* as we have seen, follows a kairomone laid down by its host. But for many other species, the supposition is that to bring itself into the ballpark, so to speak, the parasite must first locate the plant on which its host feeds. Once there, it has narrowed its search considerably.

The cotton plant gives off odors that may provide a parasite with clues. Or clues may even be supplied by other organisms that are an integral part of the cotton community. For example, researchers have found that gases produced by a fungus in rotting peaches attracted a laval parasite of the tephritid fruit fly—whether or not the fly was present. Similarly, parasites of pine bark beetles or pests on sugar cane have been attracted to the immediate area when feeding by their hosts released fragrant terpenes or sugars from the plants.

"We are identifying the chemicals in cotton that attract parasites," Vinson said. "Once we have done that, we have a chance to manipulate parasites for more effective pest control. If we find a chemical in cotton that attracts a parasite of *Heliothis,* maybe we can increase that parasite's ability to locate the plant quickly. We may be able to make an extract of that chemical, increasing its concentration, and either breed it into cotton or spray it onto individual plants. A former student of mine has already been successful in attracting parasites by the latter method."

In related work, Vinson and his students have prepared parasites in advance to locate suitable environments more quickly. They conditioned the insects to an extract from cotton by spraying the substance on a surface where they would come into contact with it in the laboratory. Upon release, the parasites made their way directly to cotton plants. But when the researchers conditioned them first to extracts from tobacco, the parasites flew to tobacco plants instead.

Ultimately, of course, the parasite must locate its host insect on the plant. "Kairomones seem to turn on the parasite to search for the host," Vinson said. "If a parasite searches a plant for a few minutes and doesn't find its host, it may fly away. The longer you can keep a parasite on a plant, the better chance it has of finding its host. We think we can hold it there longer with extracts of its host's distinctive chemicals."

Vinson has learned that some parasites are attracted by kairomones in the frass or excrement of the host. "The material can be removed by soaking or macerating the frass in an appropriate solvent," he explained. "A wide range of chemicals act as kairomones, so a series of chemicals of increasing polarities can be used to extract the compounds."

Another area of research that interests Vinson is the augmentative approach to biological control. He believes the key to successful control on row crops is the entomologists' ability to reinforce populations of natural enemies that are already in the field. Augmentation, however, demands massive springtime releases that will squelch pest outbreaks without insecticides at the very beginning. A major obstacle to this approach is the difficulty of rearing enough parasites. Entomologists need great numbers of the host insect in the laboratory to feed their "crop." Vinson has experimented with the common egg parasite *Trichogramma,* inducing it to lay thousands of eggs in a nutritious artificial media doctored by kairomones to pass them off on the parasite as the genuine eggs of its host.

"Rearing parasites is a roundabout process to begin

with," Vinson explained. "You can't just remove the eggs from an adult female parasite, as you could do with, let's say, a salmon, and then fertilize them. The eggs of these parasites will hatch only when activated by pressure that is exerted on them as they pass down the canal of the ovipositor. So, as we did with *Trichogramma,* we try to use a chemical to induce parasites to lay eggs in artificial situations, collect them, and then rear them. We aren't at that stage yet where we can apply it practically, but we're making progress."

Vinson foresees a whole new industry devoted to "insect husbandry," in which companies will breed new strains of beneficial insects with such desirable characteristics as fecundity and a superior searching ability. He has already lent his expertise to one such commercial venture.

"Here at Texas A&M I've worked on one of the boll weevil parasites, *Bracon mellitor,*" he said, "but it's not a good one. When the weevil population is low, this parasite doesn't search it out, but attacks other species that are easier to find. Anyway, there's a company in this area that rears a variety of parasites for sale. They wanted to rear *Bracon mellitor* as a general parasite against pests, but they found it expensive to rear the boll weevil for a host in the lab. They wanted a cheaper host with no serious diapause problems and so forth, and they had a good one in the wax moth [*Galleria mellonella*].

"But another problem emerged: The wax moth turned out to be one of the few insects *Bracon mellitor* won't attack! So they came to us for advice and I suggested they extract a kairomone from the boll weevil and spray it on the wax moth larvae. It worked, so you see this can have a practical application."

A practical application is the goal of all the work now being done in the Integrated Pest Management Program at

Texas A&M. Perry Adkisson and the others have worked on the belief that the program would be a futile academic exercise unless its primary goal was to enable cotton growers to earn a profit. The link must remain strong between the researchers at College Station and the growers in the field.

A tangible expression of that link is the Texas Pest Management Association, a farmer-run organization founded in 1977. The membership has followed a course of integrated management, relying heavily on a scouting system (toward which the growers contribute one-quarter of a million dollars a year) to keep abreast of insect problems in the field. Raymond E. Frisbie, Integrated Pest Management Coordinator, serves as a technical advisor to the board and attends all its meetings, but most of the day-to-day decisions are made on the scene.

"On the local level the work is coordinated by an extension agent for pest management who is trained in integrated pest management," Frisbie said. "He is responsible for a two-county area, as a rule, and he must live in the community year-round. During the winter he puts on education programs for the farmers and begins to line up his scouts— agricultural students, high school teachers, and others— who will monitor the fields for pest outbreaks during the growing season. In fact, we've found that the best scouts are the wives of the cotton growers themselves. They're knowledgeable and conscientious and they're not about to exaggerate the seriousness of an outbreak because they appreciate the necessity of keeping chemical costs down."

In spring, the extension agent for pest management conducts a week of classes designed to familiarize the scouts with their duties in the field. When the growing season begins, they scout the fields on a regular basis, sampling the plants for insect densities and plant damage. (The Pest Management Association pays their salaries and travel expenses from a fund to which all the members contribute.) The days of the "washday program" are past. Few Texas cotton farmers spray by the calendar anymore; those that use chemicals at all generally wait until the scouting reports show that a major outbreak of pests is underway.

"The program works," Frisbie insisted. "During the mid-1960s, Texas farmers were putting 20 million pounds of insecticides on cotton. Today the total is about 2 million pounds. With the new short-season varieties, the grower's yield is down, but so is his investment in chemicals, fertilizers, and water. His *net* income has gone up—sometimes as much as two and a half times an acre! And that's what he's in business for."

Knox Walker, who has lived all his life at College Station in the heart of the cotton belt, regards the shifting patterns with the eye of a man bemused by the human comedy. He watched the desperate race between the maturing plant and the boll weevil. He watched the technocrats' attempt to submerge the weevil in a sea of chemicals. Today, he plays a role in reviving the age-old strategy of fighting nature with nature, but he points out that the integrated pest management program designed by Texas A&M receives vital help from the periodic weevil-killing droughts that strike the state's newer cotton-growing areas in the High Plains; the wetter regions of the American Southeast would need equally imaginative, yet vastly different, designs if the program were to succeed there.

"We're not any brighter over here in Texas than anywhere else," he said. "It's just that environmental reality has been thrust on us."

20

POLITICS VERSUS PROGRESS
The Learning Curve in California

SOMEONE HAS REMARKED THAT OF THE VARIOUS WAYS OF MAK-ing history, "the surest way is to do harm to others." The lo-cust and the boll weevil have chewed their way into the history books, but no other insect, year after year, has made itself so well known to the human consciousness as the mos-quito.

There are perhaps 2,500 species of mosquitoes known to science, so that the tribe girdles the earth and extends even beyond the Arctic Circle. They are true flies, in the order Diptera and the family Culicidae, with a single pair of wings and slender legs. The hum, sometimes rising as they approach to an unnerving whine and caused by the vibrating of their wings, is sufficient to impress us with their presence, especially when that "song" may be the prelude to a bite. They have tried the patience of philosophers, and centuries ago, wise men in India wondered whether the protective umbrella of non-violence should be extended to cover mosquitoes.

But a mosquito's bite can be more than a minor irritation. It may transmit viruses and other pathogens that cause diseases such as malaria, yellow fever, and elephantiasis. Mosquitoes, in fact, bite an enormous range of living things, including most warm-blooded animals and even amphibians.

Only female mosquitoes possess the piercing and sucking mouthparts needed for biting. Without a blood meal, the females of most species cannot lay viable eggs. Male mosquitoes are harmless creatures that may suck plant juices briefly, but die soon after mating. The occasional report of a biting male has almost always, upon investigation, been traced to a hermaphrodite.

The female lays her eggs in shallow water, often in temporary pools such as those that accumulate in rotting stumps, discarded cans, and automobile tires—even in a pet's water dish or in damp ground that will be covered with water in the future. Eggs laid in water hatch quickly and the larvae remain there, feeding mostly on plant materials until they have pupated and emerged into the air as winged adults. Eggs laid elsewhere cannot hatch until the area is flooded; they may lie dormant through several dry years until the rains arrive or humans inadvertently supply the water needed to support the larvae's aquatic life.

The most practical time to try to control mosquitoes in the open environment is during their aquatic stage. Pest control workers for many years have sprayed chemical in-

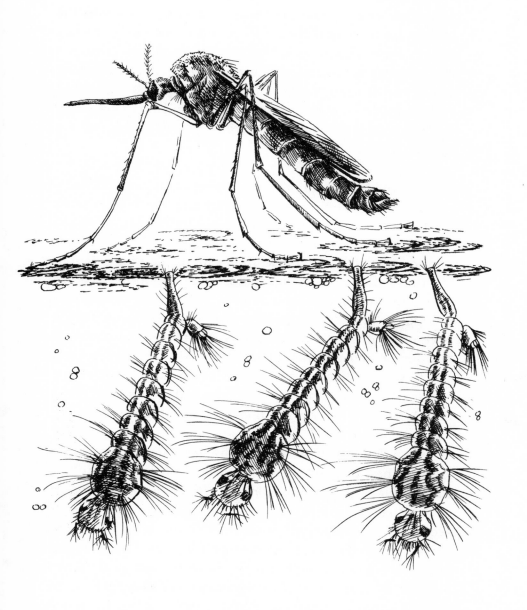

The culex mosquito and its underwater larvae.

secticides over marshes or bodies of water, creating pro-
found environmental disturbances in the process. Even projects to control adult mosquitoes with chemicals indoors in
malaria-plagued regions came a cropper when these pests
acquired resistance to the sprays.

In much of the world, the mosquito remains a monumental nuisance as well as a serious public health menace.

The University of California's Division of Biological Control
has its headquarters on what is known as the Gill Tract, behind a wire fence on busy and sometimes seedy San Pablo
Avenue in Albany, across the city line from Berkeley. The
tract consists of a few plots, where plants can be grown for
use in the division's experimental work, and a spread of low
buildings and greenhouses honeycombed with laboratories.
That is the physical installation. Yet in talking to staff members, one picks up a hint of something more. There is a certain fervor and urgency about the work, never very overt,
and frequent references to past remarks or achievements of
"Van." The spirit and the style are both reminders of the
man who was, for a time, the division's chairman, Robert
van den Bosch.

Van den Bosch was one of the great foreign collectors.
Around the Gill Tract they still speak of his bringing to heel
the walnut aphid (*Chromophis juglandicola*), an Old World
insect that became a pest in California's walnut-growing
areas, despite repeated control attempts with chemicals
(and the usual environmental upsets). Most entomologists
believed the aphid to be invulnerable to biological control
because it increases to such high numbers so quickly in the
spring. Van den Bosch wondered if the problem lay in the
complex of natural enemies relied on until then, for none of
them were host specific. Accordingly, he searched European
walnut groves in 1959 and finally found a parasitic wasp,

Trioxys pallidus, in southern France. This insect was found to be specific for the walnut aphid and very soon had the pest under control in southern California's coastal plain.

The parasite's success, however, was only regional. The aphid remained a pest in the very hot, dry summers of central and northern California. Van den Bosch remembered that he had seen *Trioxys pallidus* on an earlier trip to Iran's central plateau, which also has an extremely hot, dry climate. He returned to Iran, collected the local strain of that species, and released it in northern and central California, where it has become the decisive element in controlling the walnut aphid. This introduction provided one of the first dramatic examples of how important it is to fit not only the parasite species, but also its climatic strain, or ecotype, to the regional populations of a wide-ranging pest.

Van den Bosch was not one of those cultists who shy from any taint of chemicals with the same horror John the Baptist showed for Salome's charms. "We need pesticides," he insisted. He believed that, under existing agricultural conditions, it would be foolhardy to abandon them entirely. Modern agriculture would have to be weaned away from them, but meanwhile they must be used judiciously as part of a broad pest control strategy. He once estimated that about 20 percent of his scientific papers dealt with the *proper* use of chemicals. Although the techniques of an integrated system of controls had been known to a few entomologists for a long time, van den Bosch's name is among those most closely identified with the modern concepts of integrated pest management. Writing in *Hilgardia* in 1959, van den Bosch and three of his colleagues (Vernon M. Stern, Ray F. Smith, and Kenneth S. Hagen) set forth the principles of integrated pest management that have worked so well when carried out conscientiously on cotton and other important crops.

Van den Bosch came to his convictions about alternate

methods of pest control from personal experience. At first, this experience was entirely enjoyable.

"Insects have always fascinated me," he once wrote. "As a tyke I constantly relieved the neighborhood gardens of their lady beetles, bumble bees, and butterflies, and cluttered the house with my prizes. Today, my happiest hours as a professional entomologist occur when I am collecting or observing insects in the field. The majority of my friends are entomologists, for the most part gentle, scholarly people who occupy themselves with the biological doings of such creatures as lady beetles, plant lice, fruit flies, mini-wasps, chiggers, wolf spiders, and similar obscure but fascinating species."

He carried this enthusiasm into his career as a foreign collector. Enthusiasm turned to unease and later to outrage, however, as he observed the abuse of California's agricultural lands by the proponents of chemical treatment. In these people's hands, integrated pest management became burdened by what one might politely call a "tactical definition," twisted to meet the concerns of the sellers of chemicals. Believing that pest control is a science and a process, van den Bosch watched in disbelief as his discipline fell under the control of those who saw it only as a substance—a chemical whose sale and application would solve all humanity's pest problems. For van den Bosch, success in dealing with those problems came to hinge on the outcome of the struggle between the ecologists and the salesmen. When he railed against those abuses, he incurred the hostility of California's agricultural establishment, and there were attempts to intimidate him within the state university system. Yet no one was ever able to shut him up.

"The idyllic world of beetles and butterflies has largely slipped away as I have become increasingly involved in the roaring pesticide controversy; a vicious, nerve-racking imbroglio that has turned my entomological niche into a veritable hornet's nest," he wrote. "What is most saddening is that . . . I have turned into a ruthless gut-fighter in a slugfest without rules or a semblance of fair play."

Those words are from his book, *The Pesticide Conspiracy,* which was published in 1978. It is a blunderbuss of a book, in which he blazes away at the people and institutions that are fighting a reasonable pest control strategy in the United States. His book's thesis is that corruption lies everywhere in the pest control field—is, in fact, its very bone and sinew. He finds the Entomological Society of American rotten with the influence of the chemical companies; its members beguiled by hospitality suites, fishing holidays, and for the more "deserving" among them, industry minigrants. In his view, chemical salesmen are "rip-off artists," their executives are "Mafia *capi,*" and agricultural commissioners are "in collusion" with the salesmen. The USDA officials spend most of their energy truckling to politicians who are themselves a part of the "pesticide Mafia," and the land-grant universities are mostly anachronisms whose viewpoint excludes all society's interests except that of the powerful agribusiness pressure groups.

It is a cynical book, yet also the cry of a scientist who has seen the corruption at first hand and been told of other seamy incidents by trusted friends. Still, he believed that the principles of science were worth fighting for, and that what he had to reveal might influence public action. Among the few heroes in his book were Dr. Allen Telford and his fellow workers in the Marin County Mosquito Abatement District who abandoned the time-honored practice of spraying the 2,000-acre Petaluma Marsh five times a year with the powerful biocide, parathion. Instead, they developed an integrated program, which included a determination of the areas in the marsh that needed no "management" of any kind. But the team's most important discovery was that the marsh's big sloughs and channels were subject to tidal flushing that automatically swept the developing pests out to sea. The problem was caused mainly by potholes, created during World War II when the marsh served as a practice bombing range. The potholes were not subject to tidal flushing. Telford and his team used a ditching machine to establish a pothole draining system, thus ridding the area of most of the mosquitoes and cutting spraying by 90 percent.

Van den Bosch died suddenly in the fall of 1978, but his name is used at the Gill Tract as if he were still a part of the scene. And, in a sense, he is.

Richard Garcia has spent much of his time in recent years working on mosquito problems. Like his former chief at the Gill Tract, van den Bosch, he believes insect problems can only be solved when their underlying causes have been explored. In summer, he is likely to be found at the Gray Lodge Wildlife Area, near Yuba City, more than 100 miles northeast of Albany. This large state refuge, which was formerly a duck hunters' club, is host to hundreds of thousands of migratory birds including sandhill cranes and Ross' geese.

"Most of the mosquito breeding sources in California today are the creations of agriculture," said Garcia, a dark-haired, compactly built man. "There are 600,000 acres of rice fields, which are the sources of two or three species of mosquitoes that are important pests. Irrigation is another source. Much of the land that is irrigated for cattle is hard pan, with poor percolation, so even when the water is drained off, small pools of warm water with lots of nutrients remain in the hoofprints made by the cattle. Because of the temporary nature of this water, there are not apt to be aquatic predators around. These areas produce *tremendous* numbers of day-biting mosquitoes.

"Duck farms and wildlife refuges make perfect breeding places, too. Parts of them, such as at Gray Lodge, are diked and flooded to create different kinds of waterfowl food, sometimes several times a year. I can go into Gray Lodge with a small dipper and dip out 300 or 400 larvae at a time. In 10 days they are ready to fly. When the area is drained, the adults lay their eggs in the mud, and as soon as it's flooded again you get a new hatch, and so on."

Garcia is concentrating on the use of a new variety of *Bacillus thuringiensis* developed by scientists in Israel and given the name *israelensis* (BTI). It harms only mosquitoes, blackflies, and a handful of very close relatives.

"It isn't really a biological control, but a microbial insecticide," Garcia said. "It works the way *B.t.* does on spruce budworm. It isn't the infection itself that causes death, but a toxin created by the spores, so what we are really doing is using the bacteria in the field to manufacture a poison that kills only mosquitoes. Industry has started to produce it commercially, because there is a huge dilution factor with the bacteria. You can grow enormous quantities in the factory, but you don't need much of it in any given situation. It has great killing power. I believe it will be the first major insect pathogen to be used on a worldwide basis."

And then Garcia was thoughtful for a moment.

"I don't believe *israelensis* is a long-term panacea," he said. "People are moving into it because they don't like to deal with complex answers, such as water management. The solution to mosquito problems, when it comes, will deal with a variety of environmental factors."

Kenneth S. Hagen, a large, gregarious man, is the author of a number of important papers with other famous names in biological control, including Paul DeBach, Ray F. Smith, and Robert van den Bosch. Among his least known projects (but one that holds tantalizing possibilities for the future) was his work on an "artificial egg."

"The idea came from observations of *Trichogramma,* a genus of parasitic wasps, trying to oviposit into 'unnatural' objects such as plant seeds, bits of glass, and even the partly dried sap from an okra plant," Hagen said. "A dozen or so years ago there was a graduate student here from India, G. F. Rajendram. We decided to select a wasp from that genus to see if we could get it to oviposit readily into paraffin droplets. They were tiny hemispheres formed of a mixture of paraffin and white petroleum jelly and enclosing different solutions, from distilled water, to mixtures of amino acids, and saline solutions."

Hagen and his student selected *Trichogramma californicum,* a species reared from the Douglas fir tussock moth. The wasps oviposited in all the droplets except the ones containing distilled water. One of the artificial eggs containing a saline solution received nearly 500 parasite eggs, a phenomenon that Hagen thinks occurred because the first eggs laid in the droplet "introduced chemicals that made the solution more attractive for further oviposition"; in other words, the saline solution had a synergistic effect with the chemicals from the parasite eggs, attracting more parasites rather than turning them away as would be likely to happen with a host egg already parasitized in nature.

The parasite eggs, of course, did not develop in the experimental solutions, nor did Hagen and his student ever get around to preparing a nutritious artificial diet to go with the bogus egg. For one thing, *Trichogramma* is not an especially popular parasite in the United States because most of its species attack a great variety of hosts.

"But with more technology and some tinkering, the

genus might become an important parasite against pests in row crops," Hagen said. "The Chinese are using those wasps now quite extensively on their big collective farms. It's very labor-intensive farming. They have a machine that grinds out these neat little eggs, probably of paraffin, or maybe of plastic—they won't let you look into the machine. But they rear *Trichogramma* in enormous numbers, and they have the labor force to keep inundating the fields with them. They don't get perfect control, of course, because in China they're willing to take a little damage. Not like here, where the big producers are after perfection. If you turned out enough *Trichogramma* to try for 99 percent control, you would eventually induce resistance against the parasite. You would get selection among the pests—such as developing thicker egg coverings that the parasite couldn't penetrate with its ovipositor."

Robert van den Bosch achieving a permanent success in the control of the walnut aphid. . . Dick Garcia helping to fit into place another weapon in our age-old struggle against the mosquito tribe . . . Ken Hagen scheming to add another effective ally to biological control's growing arsenal. Work at the Gill Tract over the years is characterized by imaginative science tempered by a nice discretion. One of the cornerstones of this science is the resolve to keep technology under tight rein. The idea is to walk the straight and narrow between chemical disruption, on the one hand, and scattershot science, on the other, for otherwise the balance will tilt toward the wrong party in that endless race between the natural enemy and its host.

21

COEXISTENCE AT HOME
Our Fields, Our Farms, Your Own Backyard

THE DRAGON HUNTERS ARE BIOLOGISTS IN THE TRUE SENSE, working within the web of life, fitting the pieces together to maintain a healthy environment for all living things while plotting to pull the "pest" (which has often increased to unnatural proportions because of some human blunder) back to its proper place in the relevant ecosystem. It must never be forgotten that the purpose of applied biological control in agriculture is economic. It has no reason for existence unless it enables the grower to make a profit. The experience of the last half century has proved that pest control by chemical insecticides simply does not work. The door is now open for alternative techniques—if the farmer can be convinced of their economic advantage.

In recent years, the popular press has reported much research into new areas of pest control that approaches high technology in its sophistication. Scientists are looking into the respective merits of juvenile hormone analogs, pheromones, and the sterilization of pests by chemicals and radiation. Although some advances, and even tangible successes (as in the screwworm campaign), have been achieved with these techniques, in general their development for field use has been both slow and expensive.

- A few years ago great hope was held out for the scientific manipulation of the hormones secreted by insects themselves. Studies at Harvard University and elsewhere had demonstrated that an insect requires these hormones at certain stages of its life if it is to mature normally; their presence at other stages interferes with its development. For instance, when scientists synthesized an analog of the hormone needed by a larva for normal growth and applied it to the insect egg, growth patterns were deranged and the egg failed to develop.

  The technique is extremely interesting and may have a place in future pest control programs. Until now, however, problems associated with costs and delivery (getting the substance to pests at the right stage of their lives in the open environment) have ruled it out as a practical alternative to chemicals.

- The use of pheromones—those chemicals an insect secretes to signal other individuals of its species—has already proved to be of practical value. Scientists have baited traps with pheromones to lure pests to where they can be destroyed or sterilized. In some programs, as in that to control the gypsy moth, pheromones are useful in trapping insects so that scientists can accurately sample populations and predict the severity or decline of an infestation.

There has also been research into ways to manipulate chemical signals to trick insects, to the farmer's advantage. For instance, scientists at Cornell University have discovered that aphids, when seized by a natural enemy, emit a warning chemical that sends other aphids fleeing from the plant. By synthesizing a stable analog of the alarm chemical, these scientists have found that other aphids will not return to the area for as long as they can detect its presence. Again, there is still a great deal of work to be done in manipulating such chemicals for practical effects.

• The sterile male technique proved its value against the screwworm, as well as the pink bollworm, but its general use is limited. The technique is expensive and is suited to only a small percentage of pest insects, mainly those that are present in comparatively small numbers. A scheme developed within the USDA to try to eradicate the cotton boll weevil by the sterile moth technique brought a storm of protest from entomologists, who realized that the weevil population would first have to be reduced to manageable proportions by a massive application of chemical insecticides throughout the South.

A beguiling offshoot of this technique is the use of low levels of radiation to introduce a disastrous genetic load within a pest population. Ross A. Nielsen, of the USDA's Bee Breeding and Stock Center in Baton

Rouge, Louisiana, irradiated greater wax moths (*Galleria mellonella*), which are serious pests in commercial beehives, and released them in infested areas. These treated moths introduced into the local population some highly deleterious recessive mutations that began to appear in the second generation. Legs or palpi fell off some of the adult moths; the larvae refused to feed, or grew slowly and molted with difficulty, so that the pest declined to an insignificant level in the surrounding region. Once more, additional research to perfect this technique over wide areas has yet to be carried out.

There is no such "castles-in-the-air" aura about classical biological control. As long ago as 1970, Paul DeBach calculated that there already had been 235 successful importations of natural enemies (each resulting in the permanent control of a serious pest) in various countries around the world. Many more cases of successful biological control have been seen since then. DeBach's data show that "the chances of obtaining some significant extent of success against any given pest species from important natural enemies are about 54 in 100," and the savings in damage prevented comes to about 30 dollars for each dollar spent. This figure compares very favorably with the benefit/cost ratio for chemical insecticides of about five to one. Even when biological control is not completely successful, it may cut pest control costs at least in half. Industry spends millions of dollars to develop a new insecticide (recent estimates exceed 4 million dollars), whereas the introduction of a natural enemy seldom costs more than a few thousand dollars.

"What are biological control's implications for me?" the home gardener is likely to ask. Conservation, as the twentieth century draws to a close, is a word more honored in rhetoric than in practice. We speak of the conservation of energy, of water, of endangered species, often without letting it influence our behavior when dealing with those resources. But for the gardener who is interested in biological insect control, it is the key word—because without the conservation of natural enemies, biological control is meaningless.

We have seen what a complex and imaginative discipline applied biological control can be in the hands of an expert. The tools and techniques needed are generally far beyond the capacity of the individual who farms or gardens on a limited scale. Yet in a larger sense, most of the tools for a biological control program are already at hand, and the techniques can be summed up in the word *conservation.* "Insects are their own worst enemies," Robert van den Bosch wrote. A good deal of the work of controlling pests in the garden can be taken care of by other insects.

The trick here is not to interfere with the parasites and predators that are already at work. Asa Fitch, the nineteenth-century entomologist, used to tell the story of a man who diligently went out and cleaned all the "old parent bugs" off his roses, and then could not understand why the aphids ran wild; it turned out that this man had mistaken ladybird beetles for adult aphids. One of the starting points, then, for the gardener interested in biological control is to learn (with the aid of a good guidebook) the identities of such effective natural enemies as ladybirds, green lacewings, assassin bugs, praying mantises, and tachinid flies. Other destroyers of pests include spiders, predacious mites, and many species of ants (although the ants seen running around on plants tending aphids or mealybugs are pests to be discouraged, too).

This is not to say that biological control for the gardener

with a small plot is simply a passive exercise. The plant-feeding insects that we consider pests are, as we have seen, hardy and adaptable organisms that are not likely to inspire complacency. Successful gardeners do all they can to create conditions that will discourage pests while encouraging their natural enemies. Contemporary books and publications on organic gardening offer good advice on planning diversified gardens in such a way as to make the individual plants less vulnerable to plant-feeding insects. At the same time, the gardener can grow—or at least refrain from banishing— a number of plants in or near the garden that are not for the "pot." These plants can provide shelter and resting places for beneficial insects, as well as nectar-producing flowers that offer an abundant food supply for parasitic wasps and other natural enemies.

All gardeners expect to lose some part of their harvest to insects. The point here is not to panic and spray to wipe out

A grower steps on the chemical treadmill.

the complex of natural enemies and trigger a worse problem than that found in the first place. The home gardener can tolerate a slightly blemished tomato, which might be culled in the market.

When an outbreak of a certain pest occurs, it can often be controlled in a small garden by picking the offenders off by hand or washing them away with a stream of water from the hose. In emergencies, the gardener may want to select one of the less harmful insecticides, such as the plant-derived substances rotenone, nicotine, or derris. Even more to the point are the extremely specific microbial insecticides developed in recent years. *Bacillus thuringiensis (B.t.)* is now registered for use against a large number of garden pests and sold under such brand names as Biotrol, Dipel, and Thuricide. (For further information on the art of biological control, the reader may want to consult Paul DeBach's *Biological Control by Natural Enemies*, which is listed in the Acknowledgments and Bibliography, or write to the International Organization of Biological Control, 1050 San Pablo Avenue, Albany, California 94706.)

Most advances in classical biological control will take place in agribusiness, as it becomes more apparent that chemical control leads the grower into a cul-de-sac. The concluding years of the twentieth century will be an era of ingenious research into techniques that will extend the protection given by natural enemies over a wider range of pest species. This work will include:

- More sophisticated research, both in North America and abroad, into the life histories of natural enemies, discovering and describing new species and matching them up with hosts that are economically important pests.

- Finding ways to mass produce natural enemies, such as the use of artificial diets to rear species of *Trichogramma* in China. In some cases, these advances in rearing will also depend upon the ability of researchers to unravel the complex reproductive strategies that are characteristic of the parasitic Hymenoptera. *Coccophagoides utilis,* an important parasite on the olive scale (*Parlatoria oleae*), was nearly abandoned in biological control before its reproductive mode was figured out: the males are produced only as hyperparasites on immature females of the same species.

- The use of genetic engineering to produce more effective natural enemies. Scientists working with fruit flies have already taken advantage of the phenomenon of "jumping genes," genetic material that moves spontaneously in and out of chromosomes. After selecting genes for favorable traits and attaching them to "jumping genes," scientists were able to inject the linked genes into the casings enclosing the embryos of defective fruit flies. When the "jumping genes" entered the embryos' chromosomes, they brought with them the genes that the scientists hoped to incorporate into the defective population. Techniques of this kind may help biological control specialists produce natural enemies equipped with more acute sensory organs, stronger ovipositors, and so forth, for better success in controlling pests.

- The manipulation of habitats to make them more attractive to natural enemies. A favorable, if inadvertent, occurrence of this kind was observed some years ago when a predacious earwig from a moist tropical region was successfully introduced into what had been an arid area of California, after irrigation canals there had created a suitable habitat for it. Scientists have already pointed out the value of altering a habitat by introducing other plants into a monoculture and thus providing shelter and nectar for adult natural enemies.

- Research to find ways to help natural enemies locate their hosts more quickly. As we saw in Chapter Nineteen, scientists are already working with the chemicals secreted by insects, such as pheremones and kairomones, to lure parasites and predators to areas where they are most likely to encounter pest species.

- The development of more selective chemical insecticides to spare at least a proportion of the natural enemies present in a sprayed field. In this case, the introduction of integrated pest management programs is already teaching pest control specialists how to work around natural enemies with their chemicals and, perhaps, gradually eliminate the use of these substances except for true emergencies.

David Pimentel, a research entomologist at Cornell, is among the scientists who believe that the use of pesticides in agriculture could be cut drastically even now, without creating a serious shortage of food in the United States. About half of the chemicals are applied to non-food crops, such as cotton and tobacco, Pimentel points out, whereas much of the rest is not needed or drifts away from the target species during aerial spraying. He also criticizes government regulations that require farmers to maintain cosmetic standards of production, so that they must spray not to protect their crop against severe damage, but rather to keep it "looking nice" for the market. The elimination of much of this wasteful spraying would conserve natural enemies and thus reduce the need for chemicals even further.

In the long run, the success of biological control (as well as other alternatives to chemical insecticides) will depend on how the farming community perceives its worth in dollars and cents. Rachel Carson's argument for taking "the other road" will prevail, not because more people come to share her concern for life, as desirable as that would be, but because chemical control no longer turns a profit.

The impressive list of successes already scored by biological control proceeds from the scientist's awareness of an insect's adaptability and the complexity of its environment; he understands the uniqueness of the individual and its capacity for adjusting to outside forces. The important choices that lie ahead for science—and society—ought to be informed ones. Workers in biological control, by asking probing questions of the natural world, offer hope that humanity can live on better terms with its only real competitors for ascendancy on this planet.

APPENDIX

Some of the firms in North America that sell insect parasites and predators or microbial insecticides, such as *Bacillus thuringiensis*.

Associates Insectary
P.O. Box 969
Santa Paula, California 93060

Beneficial Insects Company
P.O. Box 323
Brownsville, California 95919

Beneficial Insects, Ltd.
  P.O. Box 154
  Banta, California 95304

Better Yield Insects
  13310 Riverside Drive
  Tecumseh, Ontario
  Canada N8N 1B2

Biogenesis, Inc.
  Route 1, Box 36
  Mathis, Texas 78368

Bio Insect Control
  1710 S. Broadway
  Plainview, Texas 79072

Bio Insect Control
  Box 9134
  Alexandria, Virginia 22304

W. Atlee Burpee Company
  Warminster, Pennsylvania 18991

California Green Lacewings
  P.O. Box 2495
  Merced, California 95340

Connecticut Valley Biological Supply Company
  Valley Road
  Southampton, Massachusetts 01073

Gurney Seed and Nursery Company
  Yankton, South Dakota 57079

King's Natural Pest Control
  224 Yost Avenue
  Spring City, Pennsylvania 19475

Lakeland Nurseries Sales
340 Poplar Street
Hanover, Pennsylvania 17331

Mellinger's Nursery
2310 W. South Range Road
North Lima, Ohio 44452

Mincemoyer's Nursery
R.D. 7, Box 379
Jackson, New Jersey 08527

Nationwide Seed and Supply
Box 91073
Louisville, Kentucky 40291

Natural Pest Controls
9397 Premier Way
Sacramento, California 95826

Peaceful Valley Farm Supply
11173 Peaceful Valley Road
Nevada City, California 95959

Pest Management Group
545 West 8th Avenue
Vancouver, British Columbia
Canada V5Z 1C6

Rincon Vitova Insectaries
P.O. Box 95
Oak View, California 93022

Spalding Laboratories
Route 2, Box 737
Arroyo Grande, California 93520

ACKNOWLEDGMENTS

IT SHOULD BE APPARENT TO THE READER THAT THIS BOOK IS NOT THE product of a single author, but of the many people—chiefly the scientists who, through their papers and books, and my interviews with them—tell their own stories here. I want to thank them and their colleagues for the generosity and patience with which they explained their work to me. They include:

Robert F. Luck, Mike Rose, and Stan Warner, Division of Biological Control, University of California, Riverside, California

Miguel A. Altieri, L. E. Caltagirone, Donald L. Dahlsten, Lowell K. Etzel, Richard Garcia, Andrew P. Gutierrez, Kenneth S. Hagen, Charles E. Kennett, and Richard L. Tassan of the Di-

309

vision of Biological Control, University of California, Berkeley, California

Thomas L. Burger, M. C. Holmes, and Vera E. Montgomery, Biological Control Satellite Facility, APHIS, U.S. Department of Agriculture, Niles, Michigan

William H. Day, Richard L. Dysart, R. W. Fuester, R. M. Hendrickson, Jr., and Paul Schaefer, Beneficial Insects Research Laboratory, ARS, U.S. Department of Agriculture, Newark, Delaware

Ronald M. Weseloh, Connecticut Agricultural Experiment Station, New Haven, Connecticut

John B. Dimond, Department of Entomology, University of Maine, Orono, Maine

D. Gordon Mott, Hampden, Maine

Bruce Francis, Passamaquoddy Indian Forestry Department, Princeton, Maine

John E. Henry, Rangeland Insect Laboratory, U.S. Department of Agriculture, Bozeman, Montana

Lloyd A. Andres, Biological Control of Weeds Laboratory, ARS, U.S. Department of Agriculture, Albany, California

Perry Lee Adkisson, Luther S. Bird, Raymond E. Frisbie, Fowden G. Maxwell, Frederick W. Plapp, Jr., Winfield L. Sterling, S. Bradleigh Vinson, and J. Knox Walker, Texas A&M University, College Station, Texas

I would also like to express my appreciation to the staff of the Bangor Public Library, Bangor, Maine, for making their extensive collection of books easily accessible and obtaining for me, on inter-library loan, the often obscure books and publications I needed to complete my research.

I am also grateful to Les Line, Editor of *Audubon,* who gave me permission to reprint some of the material that originally appeared in that magazine.

BIBLIOGRAPHY

LIKE ANYONE ELSE WHO WRITES ABOUT BIOLOGICAL CONTROL, I RELIED heavily on the classic technical book in this field, *Biological Control of Insect Pests and Weeds,* edited by Paul DeBach (Rheinhold Publishing Corp., New York, 1964). Also invaluable was *Biological Control by Natural Enemies* by Paul DeBach (Cambridge University Press, London and New York, 1974). Other general books on entomology and biological control that I found extremely helpful were:

CLAUSEN, C. P. *Entomophagous Insects.* New York: McGraw-Hill, 1940.

BIBLIOGRAPHY

DETHIER, V. G. *Man's Plague? Insects and Agriculture.* Princeton, N.J.: The Darwin Press, 1976.

FARVAR, M. T., and J. P. MILTON, eds. *The Careless Technology.* Garden City, N.Y.: Natural History Press, 1972.

HUFFAKER, C. B., ed. *Biological Control.* New York: Plenum Press, 1971.

KILGORE, W. W., and R. L. DOUTT, eds. *Pest Control: Biological, Physical and Selected Chemical Methods.* New York: Academic Press, 1967.

VAN DEN BOSCH, ROBERT, P. S. MESSENGER, and A. P. GUTIERREZ. *An Introduction to Biological Control.* New York/London: Plenum Press, 1982.

Following is a selected bibliography for those readers who wish to pursue in greater detail the subjects discussed in each chapter.

INTRODUCTION

CARSON, RACHEL. *Silent Spring.* Boston: Houghton Mifflin, 1962.

DEBACH, PAUL, ed. *Biological Control of Insect Pests and Weeds.* New York: Reinhold, 1964.

DETHIER, V. G. *Man's Plague? Insects and Agriculture.* Princeton, N.J.: The Darwin Press, 1976.

DUNLAP, THOMAS R. *DDT: Scientists, Citizens, and Public Policy.* Princeton, N.J.: Princeton University Press, 1981.

GRAHAM, FRANK, JR. *Since Silent Spring.* Boston: Houghton Mifflin, 1970.

HOWARD, L. O. *History of Applied Entomology.* Washington, D.C.: Smithsonian Institution, 1930.

PERKINS, JOHN H. *Insects, Experts, and the Insecticide Crisis.* New York/London: Plenum Press, 1982.

CHAPTER I

DEBACH, PAUL, ed. *Biological Control of Insect Pests and Weeds.* New York: Reinhold, 1964.

HOWARD, L. O. *History of Applied Entomology.* Washington, D.C.: Smithsonian Institution, 1930.

KOEBELE, ALBERT. Report of a trip to Australia to investigate the natural enemies of the fluted scale. *U.S. Department of Agriculture Bureau of Entomology Bulletin* 21 (1890): 1–32.

CHAPTER 2

WOGLUM, R.S. Report of a trip to India and the Orient in search of the natural enemies of the citrus white fly. *U.S. Department of Agriculture Bureau of Entomology Bulletin* 120 (1913): 58 pp.

CHAPTER 3

DEBACH, PAUL, and C. B. HUFFAKER. "Experimental techniques for evaluation of the effectiveness of natural enemies." In *Biological Control,* edited by C. B. Huffaker, 113–140. New York: Plenum Press, 1971.

DEBACH, PAUL. *Biological Control by Natural Enemies.* London: Cambridge University Press, 1974.

ROSE, MIKE, and PAUL DEBACH. A native parasite of the bayberry whitefly. *Citrograph* 67 (1982): 272–276.

CHAPTER 4

CARSON, RACHEL. *Silent Spring.* Boston: Houghton Mifflin, 1962.

DREA, J. J., and R. W. FUESTER. Larval and pupal parasites of *Lymantria dispar* and notes on parasites of other *Lymantriidae* in Poland 1975. *Entomophaga* 24 (1979): 319–327.

U.S. Department of Agriculture (U.S. Forest Service): Abundant publications and leaflets, constantly revised or updated.

CHAPTER 5

VAN DEN BOSCH, ROBERT, P. S. MESSENGER, and A. P. GUTIERREZ. *An Introduction to Biological Control.* New York/London: Plenum Press, 1982.

WESELOH, R. M. Reduced effectiveness of the gypsy moth parasite, *Apanteles melanoscelus,* in Connecticut due to poor seasonal synchronization with its host. *Environmental Entomology* 5 (1976): 743–746.

WESELOH, R. M. Behavioral responses to the parasite, *Apanteles melanoscelus,* to gypsy moth silk. *Environmental Entomology* 5 (1976): 1128–1132.

WESELOH, R. M. Seasonal and spatial mortality patterns of *Apanteles melanoscelus* due to predators and gypsy moth hyperparasites. *Environmental Entomology* 7 (1978): 662–665.

WESELOH, R. M., and J. F. ANDERSON. Releases of *Brachymeria lasus* and *Coccygomimus disparis,* two exotic gypsy moth parasitoids, in Connecticut: Habitat preference and overwintering potential. *Annals of Entomology* 75 (1982): 46–50.

WESELOH, R. M. Implications of tree microhabitat preferences of *Compsilura concinnata* (Diptera: tachinidae) for its effectiveness as a gypsy moth parasitoid. *The Canadian Entomologist* 114 (1982): 617–622.

WESELOH, R. M., and T. G. ANDREADIS. Possible mechanism for synergism between *Bacillus thuringiensis* and the gypsy moth parasitoid, *Apanteles melanoscelus. Annals of the Entomological Society of America* 75 (1982): 435–438.

WESELOH, R. M., et al. Field confirmation of a mechanism causing synergism between *Bacillus thuringiensis* and the gypsy moth parasitoid, *Apanteles melanoscelus. Journal of Invertebrate Pathology* 41 (1983): 99–103.

CHAPTERS 6 and 7

DYSART, R. J., H. L. MALTBY, and M. H. BRUNSON. Larval parasites of *Oulema melanopus* in Europe and their colonization in the United States. *Entomophaga* 18 (1973): 133–167.

DYSART, R. J. Distribution of *Anaphes flavipes* in Europe and sources of its importation into the United States. *Entomophaga* 16 (1971): 445–452.

MALTBY, H. L., T. L. BURGER, G. E. MOOREHEAD, and V. E. MONTGOMERY. A new record of a *Trichogramma* species parasitizing the cereal leaf beetle. *Journal of Economic Entomology* 62 (1969): 1157–1158.

MALTBY, H. L., F. W. STEHR, R. C. ANDERSON, G. E. MOOREHEAD, L. C. BARTON, and J. D. PASCHKE. Establishment in the United States of *Anaphes flavipes,* an egg parasite of the cereal leaf beetle. *Journal of Economic Entomology* 64 (1971): 693–697.

MALTBY, H. L., T. L. BURGER, M. C. HOLMES, and P. R. DEWITT. The use of an unnatural host, *Lema trilineata trivittata,* for rearing the exotic egg parasite *Anaphes flavipes. Annals of the Entomological Society of America* 66 (1973): 298–301.

MONTGOMERY, V. E., and P. R. DEWITT. Morphological differences among immature stages of three genera of exotic larval parasitoids attacking the cereal leaf beetle in the United States. *Annals of the Entomological Society of America* 68 (1975): 574–578.

STEHR, F. W., P. S. GAGE, T. L. BURGER, and V. E. MONTGOMERY. Establishment in the United States of *Lemophagus curtus,* a larval parasitoid of the cereal leaf beetle. *Environmental Entomology* 3 (1974): 453–454.

CHAPTER 8

DONALDSON, P. R., W. S. MOORE, C. S. KOEHLER, and J. L. JOOS. Scales threaten ice plant in Bay Area. *California Agriculture* 32 (1978): 4–7.

FRANKIE, GORDON, and J. O. WASHBURN. Ecology and biological control of ice plant scale, *Pulvinaria,* in California. (Interim report for period 1 for July 1978 to 30 June 1981, University of California, Berkeley).

TASSAN, R. L., K. S. HAGEN, and D. V. CASSIDY. Imported natural enemies against ice plant scales in California. *California Agriculture* 36 (1982): 16–17.

CHAPTERS 9 AND 10

BUSH, G. L., R. W. NECK, and G. B. KITTO. Screwworm eradication: Inadvertent selection for noncompetitive ecotypes during mass rearing. *Science* 193 (1976): 491–493.

KNIPLING, E. F. The eradication of the screwworm fly. *Scientific American* 203 (1960): 4–48.

PERKINS, J. H. *Insects, Experts, and the Insecticide Crisis.* New York: Plenum Press, 1982.

RICHARDSON, R. H. *The Screwworm Problem: Evolution of Resistance to Biological Control.* Austin/London: University of Texas Press, 1978.

RICHARDSON, R. H., J. R. ELLISON, and W. W. AVERHOFF. Autocidal control of screwworms in North America. *Science* 215 (1982): 361–370.

RICHARDSON, R. H., et al. Letters. Mating types in screwworm populations? *Science* 218 (1982): 1143–1145.

CHAPTER I I

DEBACH, PAUL, DAVID ROSEN, and C. E. KENNETT. "Biological controls of coccids by introduced natural enemies." Ch. 7. In *Biological Control,* edited by C. B. Huffaker. New York: Plenum Press.

DEBACH, PAUL. *Biological Control by Natural Enemies.* London: Cambridge University Press, 1974.

KFIR, RAMI, and R. F. LUCK. Effects of constant and variable temperature extremes on sex ratio and progeny production by *Aphytis melinus* and *A. lingnanensis. Ecological Entomology* 4 (1979): 335–344.

LUCK, R. F., ROBERT VAN DEN BOSCH, and RICHARD GARCIA. Chemical insect control, a troubled pest management strategy. *Bioscience* 27 (1977): 606–611.

LUCK, R. F., HAGGAI PODOLER, and RAMI KFIR. Host selection and egg allocation behavior by *Aphytis melinus* and *A. lingnanensis:* Comparison of two facultatively gregarious parasitoids. *Ecological Entomology* 7 (1982): 397–408.

CHAPTER 12

DREA, J. J. "The European parasite laboratory: Sixty years of foreign exploration." In *Biological Control in Crop Production,* edited by G. C. Papavizas, 107–120. Totowa, N.J.: Allanheld, Osmun, 1981.

HENDRICKSON, R. M., JR., and S. E. BARTH. Biology of alfalfa blotch leaf miner. *Annals of the Entomological Society of America* 71 (1978): 295–298.

HENDRICKSON, R. M., JR., and S. E. BARTH. Effectiveness of native parasites against *Agromyza frontella*, an introduced pest of alfalfa. *New York Entomological Society* 87 (1979): 85–90.

HENDRICKSON, R. M., JR., and S. E. BARTH. Introduced parasites of *Agromyza frontella* in the USA. *New York Entomological Society* 87 (1979): 167–174.

HENDRICKSON, R. M., JR. Biological control of alfalfa blotch leaf miner in Delaware. In press.

CHAPTER 13

DAY, W. H. "Biological control of the alfalfa weevil in the northeastern United States." In *Biological Control in Crop Production,* edited by G. C. Papavizas, 361–374. Totowa, N.J.: Allanheld, Osmun, 1981.

DYSART, R. J., and W. H. DAY. The release and recovery of introduced parasites of the alfalfa weevil in eastern North America. *U.S. Department of Agriculture Production Research Report* No. 167, 1976.

CHAPTER 14

DIMOND, J. B., and O. N. MORRIS. "Microbial and other biological control." In *U.S. Department of Agriculture Manual,* in press.

HULME, M. A., T. J. ENNIS, and A. LAVALLÉE. Current status of *Bacillus thuringiensis* for spruce budworm control. *The Forestry Chronicle* 59 (April 1983): 58–61.

U.S. DEPARTMENT OF AGRICULTURE. Final programmatic environmental impact statement: Proposed cooperative 5-year (1981–1985) spruce budworm management program for Maine. U.S. Forest Service, Northeastern area, 1981.

CHAPTER 15

ANDRES, L. A. Integrating weed biological control agents into a pest management program. *Weed Science* (Suppl.) 30 (1982): 25–30.

CENTER, T. D. The water hyacinth weevils. *Aquatics* 4 (1982): 8, 16–19.

DELOACH, C. J. *Neochetina bruchi,* a biological control agent of water hyacinth. Host specificity in Argentina. *Annals of the Entomological Society of America* 69 (1976): 635–642.

EISNER, THOMAS, et al. Defensive use by an insect of a plant resin. *Science* 184 (1974): 996–999.

MAUGH, T. H., II. Exploring plant resistance to insects. *Science* 216 (1982): 722–723.

NEW YORK STATE COLLEGE OF AGRICULTURE AND LIFE SCIENCES. Nature's defenses await exploitation (Paul Feeny). Press release, February 17, 1977.

CHAPTER 16

BARON, STANLEY. *The Desert Locust.* London: Eyre Methuen 1972.

HENRY, J. E. Epizootiology of infections by *Nosema locustae* in grasshoppers. *Acrida* 1 (1972): 111–120.

HENRY, J. E. Natural and applied control of insects by protozoa. *Annual Review of Entomology* 26 (1981): 49–73.

HENRY, J. E., and E. A. OMA. Pest control by *Nosema locustae,* a pathogen of grasshoppers and crickets. In *Microbial Control of Pests and Plant Diseases, 1970–1980,* edited by H. D. Burges, 573–585. New York: Academic Press, 1981.

ONSAGER, J. A., N. E. REES, J. E. HENRY, and R. W FOSTER. Integration of bait formulations of *Nosema locustae* and carbaryl for control of rangeland grasshoppers. *Journal of Economic Entomology* 74 (1981): 183–187.

PARKER, J. R. "Grasshoppers." In *Insects. The Yearbook of Agriculture,* 595–605. Washington, D.C.: U.S. Department of Agriculture, 1952.

PARKER, J. R., and R. V. CONNIN. Grasshoppers: their habits and damage. U.S. Department of Agriculture: *Agriculture Information Bulletin,* no. 287 (1964): 28 pp.

CHAPTER 17

PERKINS, J. H. *Insects, Experts, and the Insecticide Crisis.* New York: Plenum Press, 1982.
WALKER, J. K. A question of cotton. Unpublished ms.

CHAPTER 18

ADKISSON, P. L., G. A. NILES, J. K. WALKER, L. S. BIRD, and H. B. SCOTT. Controlling cotton's insect pests: A new system. *Science* 216 (1982): 19–22.
BIRD, L. S. The MAR (multi-adversity resistance) system for genetic improvement of cotton. *Plant Disease* 66 (1982): 172–176.
PLAPP, F. W. Insecticides useful in IPM programs: Some theories and preliminary experiments. Unpublished paper.
PLAPP, F. W., and S. B. VINSON. Comparative toxicities of some insecticides to the tobacco budworm and its ichneumonid parasite, *Campoletis sonorensis. Environmental Entomology* 6 (1977): 381–384.
PLAPP, F. W., and D. L. BULL. Toxicity and selectivity of some insecticides to *Chrysopa carnea,* a predator of the tobacco budworm. *Environmental Entomology* 7 (1978): 431–434.
RAJAKULENDRAN, S. V., and F. W. PLAPP. Comparative toxicities of five synthetic pyrethroids to the tobacco budworm and ichneumonid parasite, *Campoletis sonorensis,* and a predator, *Chrysopa carnea. Journal of Economic Entomology* 75 (1982): 769–771.
REYNOLDS, H. T., P. L. ADKISSON, R. F. SMITH, and R. E. FRISBIE. "Cotton insect pest management." In *Introduction to Insect Pest Management,* edited by R. Metcalf and W. Luckmann, 375–441. New York: Wiley, 1982.
WALKER, J. K. "Earliness in cotton and escape from the boll weevil." In *Biology and Breeding for Resistance to Arthropods*

and Pathogens in Agricultural Plants. MP-1451 (1980). Texas Agricultural Experiment Station, Texas A&M University.

WALKER, J. K., and G. A. NILES. Population dynamics of the boll weevil and modified cotton types. *Texas Agricultural Experiment Station Bulletin 1109.* 14 pp., 1971.

WALKER, J. K., R. E. FRISBIE, and G. A. NILES. A changing perspective: *Heliothis* in short-season cottons in Texas. *ESA Bulletin* 24 (1978): 385–391.

CHAPTER 19

AGNEW, C. W., and W. L. STERLING. Predation rates of the red imported fire ant on eggs of the tobacco budworm. *Protection Ecology* 4 (1982): 151–158.

AGNEW, C. W., W. L. STERLING, and D. A. DEAN. Influence of cotton nectar on red imported fire ants and other predators. *Environmental Entomology* 11 (1982): 629–634.

CASEY, J. E., R. D. LACEWELL, and W. L. STERLING. An example of economically feasible opportunities for reducing pesticide use in commercial agriculture (1975). *Journal of Environmental Quality* 4 (1975): 60–64.

JONES, DAVY, and W. L. STERLING. Manipulation of red imported fire ants in a trap crop for boll weevil suppression. *Environmental Entomology* 8 (1979): 1073–1077.

STERLING, W. L. Imported fire ant . . . may wear a gray hat. *Texas Agricultural Progress* 24 (1978): 19–20.

TEXAS COTTON AND SORGHUM REGIONAL WORKING GROUP. Present and future pest management strategies in the control of cotton and sorghum pests in Texas (a part of a study, Alternative Pest Management Strategies in Food Production, submitted to the Office of Technology Assessment, Congress of the United States, Washington, D.C., 1979).

VINSON, S. B. "Behavioral chemicals in the augmentation of natural enemies." In *Biological Control by Augmentation of Natural Enemies,* edited by R. L. Ridgway and S. B. Vinson, 237–279. New York: Plenum Press, 1977.

CHAPTER 20

ALTIERI, M. A. Weeds may augment biological control of insects. *California Agriculture* (May–June, 1981): 22–24.

BATES, MARSTON. *The Natural History of Mosquitoes.* New York: Macmillan, 1949.

CALTAGIRONE, L. E. Landmark examples in classical biological control. *Annual Review of Entomology* 26 (1981): 213–232.

GARCIA, RICHARD, BARBARA DESROCHES, and WILLIAM TOZAR. Studies on *Bacillus thuringiensis* var. israelensis against mosquito larvae and other organisms. *Proceedings of the California Mosquito Control Association* (in press).

GARCIA, RICHARD, B. A. FEDERICI, I. M. HALL, M. S. MULLA, and C. H. SCHAEFER. BTI—a potent new biological weapon. *California Agriculture* (March 1980): 18–19.

RAJENDRAM, G. F., and K. S. HAGEN. *Trichogramma* oviposition into artificial substrates. *Environmental Entomology* 3 (1973): 399–401.

VAN DEN BOSCH, ROBERT. *The Pesticide Conspiracy.* Garden City, N.Y.: Doubleday, 1978.

INDEX